LARGE-EDDY SIMULATIONS OF TURBULENCE

Large-Eddy Simulations of Turbulence is an ideal introduction for people new to large-eddy simulation (LES), direct numerical simulation, and Reynolds-averaged Navier–Stokes simulation and makes an excellent reference for researchers. Of particular interest in the text is the detailed discussion in Chapter 2 of vorticity, pressure, and the velocity gradient tensor, which are quantities useful for probing the results of a simulation – particularly when looking for coherent vortices and coherent structures. Chapters 4 and 5 feature an in-depth discussion of spectral subgrid-scale modeling. Although physical-space models are generally more readily applied, spectral models give insight into the requirements and limitations in subgrid-scale modeling and backscattering. A third special feature is the detailed discussion in Chapter 7 of LES of compressible flows – a topic previously accessible only in articles scattered throughout the literature. This will be of interest to those dealing with supersonic flows, combustion, astrophysics, and other related topics. Chapter 8 focuses on geophysical fluid dynamics with emphasis on rotating stratified shear flows. Interesting applications of LES to storm formation are given in particular.

Marcel Lesieur, Olivier Métais, and Pierre Comte form the nucleus of the Grenoble Equipe Modélisation et Simulation de la Turbulence (the Grenoble team for modeling and simulating turbulence) and play an important role in the development of subgrid-scale modeling of turbulent flows required for large-eddy simulation and in the implementation of large-eddy simulation methodology in research and applications. They were responsible for early research on spectral subgrid-scale closure and the use of the closure approach in developing the physical-space structure-function model. More recently they have made significant contributions to the development of modeling for compressible turbulent flows.

Large-Eddy Simulations of Turbulence

MARCEL LESIEUR

Professor, National Polytechnic Institute, Grenoble

OLIVIER METAIS

Professor, National Polytechnic Institute, Grenoble

PIERRE COMTE

Professor, Louis Pasteur University, Strasbourg

CAMBRIDGE
UNIVERSITY PRESS

CAMBRIDGE UNIVERSITY PRESS
Cambridge, New York, Melbourne, Madrid, Cape Town, Singapore, São Paulo

Cambridge University Press
40 West 20th Street, New York, NY 10011-4211, USA

www.cambridge.org
Information on this title: www.cambridge.org/9780521781244

First published 2005

Printed in the United States of America

A catalog record for this publication is available from the British Library.

Library of Congress Cataloging in Publication Data

Lesieur, Marcel.
Large-eddy simulation of turbulence / Marcel Lesieur, Olivier Mitais, Pierre Comte.
    p.   cm.
Includes bibliographical references and index.
ISBN 0-521-78124-8 (hardback)
1. Turbulence – Mathematical models.   2. Eddies – Mathematical models.
I. Mitais, O. (Olivier)   II. Comte, P. (Pierre)   III. Title.
TA357.5.T87L47   2005
532′.0527′015118 – dc22   2005010537

ISBN-13   978-0-521-78124-4 hardback
ISBN-10   0-521-78124-8 hardback

Contents

Preface

In 1949, in an unpublished report to the U.S. Office of Naval Research, John von Neumann remarked of turbulence that

> the great importance of turbulence requires no further emphasis. Turbulence undoubtedly represents a central principle for many parts of physics, and a thorough understanding of its properties must be expected to lead to advances in many fields. . . . [T]urbulence represents per se an important principle in physical theory and in pure mathematics. . . . These considerations justify the view that a considerable effort towards a detailed understanding of the mechanisms of turbulence is called for. . . .[1]

Few people today would disagree with these comments on the importance of understanding turbulence and, as implied, of its prediction. And, although the turbulence problem has still yet to be "solved," our understanding of turbulence has significantly advanced since that time; this progress has come through a combination of theoretical studies, often ingenious experiments, and judicious numerical simulations. In addition, from this understanding, our ability to predict, or at least to model, turbulence has greatly improved; methods to predict turbulent flows using large-eddy simulation (LES) are the main focus of the present book.

The impact of von Neumann is still felt today in the prediction of turbulent flows, both in his work on numerical methods and in the people and the research he has influenced. The genesis of the method of large-eddy simulation (or possibly more appropriately, "simulation des grandes échelles") was in the early 1960s with the research of Joe Smagorinsky. At the time, Smagorinsky was working in von Neumann's group at Princeton, developing modeling for dissipation and diffusion in numerical weather prediction. Doug Lilly, who later worked with Smagorinsky, realized the potential for simulating turbulent flows of Smagorinky's modeling work. When Lilly joined the National Center

---

[1] Quote provided by Russell J. Donnelly.

for Atmospheric Research (NCAR), he encouraged NCAR's Jim Deardorff to pursue this line of research; Deardorff later completed the first series of LES, publishing his results in several important papers in the early 1970s. At that time at NCAR, Doug Lilly, Chuck Leith, Jim Deardorff, and later Jack Herring established a most stimulating environment for turbulence research. In addition to these first large-eddy simulations and other research, such as on the parameterization of boundary layer turbulence and studies of clear-air turbulence, the first direct numerical simulations were carried out at NCAR in that time period by Steve Orszag and Stu Patterson.

Research on large-eddy simulation is increasing rapidly as this methodology takes its place as a valuable numerical simulation tool along with direct numerical simulation and Reynolds-averaged Navier–Stokes simulation. As an example of this, a very informal survey using the Science Citation Index indicates that the number of archival papers with "large-eddy simulation" in their titles has increased almost geometrically in the past decade or so from 11 in 1990, to 25 in 1995, to 51 in 2000, and to 95 in 2003. In addition, with continuing improvements in numerical methods and also in subgrid modeling of turbulence, large-eddy simulation is being utilized more and more in applications. This can be seen, for example, in its implementation in most of the commercially available fluid dynamics codes. Undoubtedly it will become a principal tool in applications in the future.

This tremendous increase in the interest in, and of use of, LES demands the availability of books that describe the theory and modeling aspects of LES and that also give detailed examples of how it has been and can be applied. Such was the task of the authors of this book.

The three authors of this book, Marcel Lesieur, Olivier Métais, and Pierre Comte, who have formed the nucleus of the Grenoble Equipe Modélisation et Simulation de la Turbulence (the Grenoble team for modeling and simulating turbulence), are eminently qualified to write such a book. They have been very active in many developments in subgrid modeling of turbulent flows required for large-eddy simulation and also in the implementation of LES methodology in research and applications. Among other things they have been responsible for some of the first research on spectral subgrid-scale closure; using some of the ideas from this closure approach they developed the physical-space structure-function model, which has received considerable attention and use; and more recently they have led in developing modeling for compressible turbulent flows. Of course the readers will find out much more about their contributions in this book.

This book contains the basic information required for both a person new to the subject of large-eddy simulation and for use as a reference for the more experienced researcher. In addition, it contains several additional items the reader may find of special importance. The first of these items is contained in

Chapter 2, where a detailed discussion is given of vorticity, pressure, and the velocity gradient tensor, which are quantities useful for probing the results of a simulation – in particular looking for coherent vortices and coherent structures. Since the approach of large-eddy simulation focuses on the large-scale motions, which are often coherent, it is important to have the appropriate tools available to examine the simulations for such features.

Another item of special importance is the in-depth discussion of spectral subgrid-scale modeling in Chapters 4 and 5. Although physical-space models are generally more readily applied, the spectral models give more insight into the requirements and limitations in subgrid-scale modeling and related issues, such as backscattering. A third special item in the book is its detailed discussion of the large-eddy simulation of compressible flows in Chapter 7 – a subject to which the authors have made important recent contributions, and information about which has, up till now, only been available in articles scattered throughout the literature. This topic will be of interest not only to those dealing with supersonic flows but also to those interested in combustion, astrophysics, and other related topics. In the final chapter the authors go back to the origins of large-eddy simulations and discuss applications to problems in geophysical fluid mechanics. The reader will become acquainted with examples of how large-eddy simulation can enable issues that are at such high Reynolds numbers that they are available only to large-eddy simulation to be addressed. Among the topics discussed are the effects of system rotation on turbulence and the generation of storms through baroclinic instabilities.

To learn a new topic, it is often best to have available examples worked out in some detail. One of the great merits of this book is that it is filled with many examples often taken from the research of the authors. These examples are supplemented by numerous animations, which are referenced at the end of the appropriate chapters and are available on the accompanying CD-ROM.

The authors have succeeded exceptionally well in providing a book that will be valuable for both the novice and the experienced user. The book will be useful as a text or reference in graduate courses on large-eddy simulation, and it should find a place on the reference shelf of both scientists and engineers who have interest in large-eddy simulation. The book should become a significant element in this rapidly developing field of turbulence simulation.

James J. Riley
June 2004

1 Introduction to LES

1.1 Book's scope

Large-eddy simulations (LESs) of turbulent flows are extremely powerful techniques consisting in the elimination of scales smaller than some scale Δx by a proper low-pass filtering to enable suitable evolution equations for the large scales to be written. The latter maintain an intense spatio-temporal variability. Large-eddy simulation (LES) poses a very difficult theoretical problem of subgrid-scale modeling, that is, how to account for small-scale dynamics in the large-scale motion equations. LES is an invaluable tool for deciphering the vortical structure of turbulence, since it allows us to capture deterministically the formation and ulterior evolution of coherent vortices and structures. It also permits the prediction of numerous statistics associated with turbulence and induced mixing. LES applies to extremely general turbulent flows (isotropic, free-shear, wall-bounded, separated, rotating, stratified, compressible, chemically reacting, multiphase, magnetohydrodynamic, etc.). LES has contributed to a blooming industrial development in the aerodynamics of cars, trains, and planes; propulsion, turbo-machinery; thermal hydraulics; acoustics; and combustion. An important application lies in the possibility of simulating systems that allow turbulence control, which will be a major source of energy savings in the future. LES also has many applications in meteorology at various scales (small scales in the turbulent boundary layer, mesoscales, and synoptic planetary scales). Use of LES will soon enable us to predict the transport and mixing of pollution. LES is used in the ocean for understanding mixing due to vertical convection and stratification and also for understanding horizontal mesoscale eddies. LES should be very useful for understanding the generation of Earth's magnetic field in the turbulent outer mantle and as a tool for studying planetary and stellar dynamics.

It is clear that the study of *large-eddy simulations of turbulence* has become a discipline by itself. This book will try to present a global and complete account of this discipline and its vigorous developments since the early 1960s and the pioneering work of Smagorinsky [269]. We will also provide various industrial and environmental applications.

Although we do not expect the reader to be an expert in fluid dynamics and turbulence, it is not the aim of the present book to give a complete account of these aspects. We will try, however, to recall in simple terms some of them while referring to the companion textbook of Lesieur [170] for the more advanced aspects or detailed derivations on these topics.

The objective of the book is twofold. The first is to present the details of many models developed in large-eddy simulations of turbulence. The second is, through examples of application, to give the reader a thorough understanding of turbulence dynamics in isotropy, mixing layers, boundary layers, and separated flows and how such a dynamics may be deeply modified by rotation, stratification, heating, and compressibility. The book contains numerous computer-generated graphics as well as a CD-ROM with movies of some flows computed with LES (isotropic turbulence, mixing layers and jets, backward-facing steps, boundary layers and channel flows, cavities at various Mach numbers, heated-channel flows, frontal cyclogenesis in the atmosphere, etc.). This interdisciplinary textbook addresses a very wide population of graduate students, researchers, and industrial engineers in the domains of mechanical, aerospace, civil, chemical, and nuclear engineering; geophysical and astrophysical fluid dynamics; physics; and applied mathematics.

In the present chapter, we recall the basis of fluid-dynamics and turbulence theory that will be used for LES. We show the limitations of direct numerical simulations in terms of practical applications at high Reynolds numbers owing to the excessive number of degrees of freedom of the system. We recall the history of LES and finish with an analysis of unpredictability effects in the framework of LES analyses.

In Chapter 2, we are mainly concerned with coherent-vortex recognition in terms of pressure and vorticity fields as well as quantities related to the velocity-gradient tensor, such as the very efficient Q and λ_2 criteria. Applications to isotropic turbulence and backward-facing steps are provided, and animations of coherent vortices are observed in both cases.

Chapter 3 presents the LES formalism in physical space with the introduction of the famous Smagorinsky model, for which we will show how the constant may be determined. We will also study the model's wall behavior, which poses serious problems. We also present a thorough description of its more recent so-called dynamic version with a dynamic recalculation of the constant by a double filtering in space.

Chapter 4 presents spectral models for LES applied to three-dimensional isotropic turbulence with the plateau-peak eddy viscosity and eddy diffusivity and the spectral-dynamic model. The chapter shows new EDQNM[1] calculations at very high Reynolds numbers with an analysis of the well-known phenomenon of kinetic energy cascade pileup before the dissipative range (the so-called bump). The chapter contains a complete infrared study of kinetic-energy and pressure spectra done both with EDQNM and LES using the spectral models. It also discusses other types of spectral eddy viscosities such as Heisenberg's and RNG-based.

Chapter 5 shows how the plateau-peak eddy-viscosity model may be applied to inhomogeneous turbulence in flows of uniform density in the particular cases of a temporal mixing layer, where it is able to reproduce a vortex structure of quasi–two-dimensional Kelvin–Helmholtz vortices stretching thin longitudinal hairpins, or dislocated Kelvin–Helmholtz vortices undergoing helical pairing, according to the quasi–two-dimensional or three-dimensional nature of the initial forcing. A thorough LES study of the plane channel using the spectral-dynamic model is carried out at various Reynolds numbers. The study is complemented by direct numerical simulation (DNS) focusing on probability density functions of various quantities, which are discussed with respect to the vortical dynamics.

Chapter 6 presents new subgrid models, such as the structure-function model and its "selective" and "filtered" versions. These models are compared with Smagorinsky's in the framework of a temporal mixing layer. They are applied to a spatially growing mixing layer, where the influences of upstream forcing and the extent of the spanwise domain are discussed. A round jet is also looked at with alternate pairings of vortex rings qualitatively similar to helical pairing in mixing layers. The jet control by upstream perturbations of varicose, helical, or flapping types is studied, with possibilities of strongly enhancing the spreading. The backstep is reconsidered statistically. Afterward a dynamic version of the structure-function model is presented. We discuss hyperviscosities as well as a mixed structure-function/hyperviscous model that parallels in physical space the spectral plateau-peak model. We also present scale-similarity and mixed models as well as some new, recent models.

Chapter 7 is devoted to LES of compressible ideal gases (neglecting gravity effects). We work in the context of density-weighted Favre filters analogous to Favre density-weighted ensemble averages. We introduce a new thermodynamic quantity, the macrotemperature, which may be related by an equation of state to a macropressure. This greatly simplifies the LES formalism for compressible flows. Afterward we discuss the compressible mixing layer both in the temporal and spatial cases. The compressible round jet is

[1] The eddy-damped quasi-normal Markovian theory (EDQNM) is a very efficient statistical model of isotropic turbulence based on two-point closures, which will be presented in more detail in Chapter 4. It also serves to determine subgrid models for spectral large-eddy simulations.

also studied both in the subsonic and supersonic cases. Jet contol by varico-flapping excitations is studied. Then various LESs of low-Mach boundary layers developing spatially upon a flat plate are presented both in the transitional and developed stages. Animations of various vortices and structures are provided. A weakly compressible channel (one side of which contains two small spanwise grooves) is also presented with animations of quasi-longitudinal vortices traveling on both sides. We recall the main features and role of longitudinal riblets equipping boats, planes, and swimming costumes and discuss the influence of compressibility. Turbulence over a square cavity and over a transonic rectangular cavity is studied. Then the structure of turbulence in the neighborhood of the European *Hermès* space shuttle at a local Mach number of 2.5 will be examined with evidence for the presence of Görtler vortices. Finally, DNS and LES of a heated square duct will be looked at. This duct may contain riblets, which increase heat transfer significantly. A curved duct with one wall heated is also studied, and Görtler vortices are recovered.

Chapter 8 is devoted to geophysical fluid dynamics with some DNS and LES of relevance for this topic. We first present a review of geophysical flows at various scales mainly for Earth's atmosphere and oceans. We determine the associated Rossby numbers. Climate issues such as global warming, the ozone hole, El Niño, and the oceanic conveyor belt are briefly discussed. Afterward we study shear flows (free and wall-bounded) of uniform density rotating about a spanwise axis. They are looked at mainly from the point of view of DNS and LES, and we show a wide universality in the dynamics of these flows. Then we present DNS and LES studies of the instability of a baroclinic jet, showing that LES permits us to capture secondary instabilities that are dissipated in DNS. We discuss possible analogies with severe storms.

1.2 Basic principles of fluid dynamics

We work within the assumption of a continuous medium whose characteristic scales of motion are several orders of magnitude larger (by a factor of 10^4 to 10^6) than the mean free path of molecules characterizing the molecular scales. Equations of fluid motion are obtained in the following way (see Batchelor [17] and Lesieur [170]). We work in a frame that may be Galilean, or in solid-body rotation of rotation vector $\vec{\Omega}$, and consider a fluid parcel (of volume δV) of size smaller than the characteristic scales in the flow. Let ρ be the density, and let $\vec{u}$ be the velocity of the parcel gravity center. One introduces the operator D/Dt, the derivative following the fluid motion, which is equal to $\partial/\partial t + \vec{u} \cdot \vec{\nabla}$ if the flow quantities are expressed in terms of a given space point $\vec{x}$ and time t (Eulerian notations). Notice that we have, respectively, for any scalar $A(\vec{x}, t)$ and vector $\vec{a}(\vec{x}, t)$

$$\frac{DA}{Dt} = \frac{\partial A}{\partial t} + \vec{u} \cdot \vec{\nabla} A, \tag{1.1}$$

$$\frac{D\vec{a}}{Dt} = \frac{\partial \vec{a}}{\partial t} + (\vec{u} \cdot \vec{\nabla})\vec{a} = \frac{\partial \vec{a}}{\partial t} + \vec{\nabla}\vec{a} \otimes \vec{u}, \tag{1.2}$$

where $\otimes$ stands for a tensorial product. The three following principles are applied to the parcel in its motion:

- conservation of mass ($\delta m = \rho \, \delta V$),
- balance of forces (Newton's first and third principles stated in 1687), and
- first principle of thermodynamics.

1.2.1 Continuity equation

The conservation of mass yields the continuity equation

$$\frac{1}{\delta m} \frac{D(\delta m)}{Dt} = \frac{1}{\rho} \frac{D\rho}{Dt} + \frac{1}{\delta V} \frac{D(\delta V)}{Dt},$$

which yields

$$\frac{1}{\rho} \frac{D\rho}{Dt} + \vec{\nabla} \cdot \vec{u} = 0. \tag{1.3}$$

The particular case of incompressibility (conservation of volumes following the fluid motion) reduces to $\vec{\nabla} \cdot \vec{u} = 0$.

1.2.2 Balance of forces

The balance of forces corresponds to the so-called Navier–Stokes equation. It is obtained by equating the "acceleration quantity" $\delta m \, D\vec{u}/Dt$ to the body forces plus the surface forces acting upon the external surface of the parcel. The body forces applied are gravity, $\delta m \, \vec{g}$, the Coriolis force (if any), $-2 \, \delta m \, \vec{\Omega} \times \vec{u}$, and other possible forces. The gravity $\vec{g}$ is irrotational and includes both the Newtonian gravity and the centrifugal force implied by the frame rotation. One assumes the existence of a stress tensor $\overline{\overline{\sigma}}$ such that the force exerted by the fluid on one side of a small surface $d\Sigma$ oriented by a normal unit vector $\vec{n}$ is given by $d\vec{f} = \overline{\overline{\sigma}} \otimes \vec{n} \, d\Sigma$. A Newtonian fluid corresponds to a stress tensor of the form

$$\sigma_{ij} = -p \, \delta_{ij} + \mu \left[\left(\frac{\partial u_i}{\partial x_j} + \frac{\partial u_j}{\partial x_i} \right) - \frac{2}{3} \vec{\nabla} \cdot \vec{u} \, \delta_{ij} \right], \tag{1.4}$$

where the pressure is defined by $p = -(1/3)\sigma_{ii}$, and μ is the dynamic viscosity coefficient. Such a definition of pressure avoids the introduction of a second viscosity coefficient. After integration of the surface forces over the surface of the fluid particle, we have

$$\frac{Du_i}{Dt} = (\vec{g} - 2\vec{\Omega} \times \vec{u})_i + \frac{1}{\rho}\frac{\partial \sigma_{ij}}{\partial x_j} \qquad (1.5)$$

or, equivalently,

$$\frac{Du_i}{Dt} = (\vec{g} - 2\vec{\Omega} \times \vec{u})_i - \frac{1}{\rho}\frac{\partial p}{\partial x_i} + \frac{1}{\rho}\frac{\partial}{\partial x_j}\mu\left[\left(\frac{\partial u_i}{\partial x_j} + \frac{\partial u_j}{\partial x_i}\right) - \frac{2}{3}\vec{\nabla}\cdot\vec{u}\,\delta_{ij}\right]. \qquad (1.6)$$

Introducing the geopotential Φ such that $\vec{g} = -\vec{\nabla}\Phi$, we can write the Navier–Stokes equation as

$$\frac{\partial \vec{u}}{\partial t} + (\vec{\omega} + 2\vec{\Omega}) \times \vec{u} = -\frac{1}{\rho}\vec{\nabla}p - \vec{\nabla}\left(\Phi + \frac{\vec{u}^2}{2}\right) + \text{vicous dissipation}, \qquad (1.7)$$

where $\vec{\omega} = \vec{\nabla} \times \vec{u}$ is the relative vorticity of the fluid (in the rotating frame) and $\vec{\omega}_a = \vec{\omega} + 2\vec{\Omega}$ is the absolute vorticity in the absolute frame. In Eq. (1.7), the viscous contribution has not been explicitly specified.

1.2.3 Thermodynamic equation

A third equation is obtained by applying the first principle of thermodynamics to the fluid parcel: The derivative of the total energy (internal, potential, and kinetic) is equal to a possible heating (or cooling) rate by some source (e.g., radiation, combustion, condensation, or evaporation of water in the atmosphere), plus the power of surface forces, plus the rate of heat exchange by molecular diffusion across the parcel surface. The latter is expressed with the aid of Fourier's law. More specifically, let e_i be the internal energy per unit mass. Then

$$\frac{De_i}{Dt} = \dot{Q} + \frac{1}{\rho}\vec{\nabla}\cdot(\lambda\vec{\nabla}T) - \frac{p}{\rho}\vec{\nabla}\cdot\vec{u} + 2\nu\left(S_{ij}S_{ij} - \frac{1}{3}S_{ii}S_{jj}\right) \qquad (1.8)$$

with $\overline{\overline{S}} = [\vec{\nabla}\vec{u} + \vec{\nabla}\vec{u}|^t]/2$, $\dot{Q}$ characterizing the forcing, and λ being the thermal conductivity. Let $h = e_i + (p/\rho)$ be the enthalpy of the fluid. From the continuity equation, we have

$$\rho\frac{D}{Dt}\left(\frac{p}{\rho}\right) = \frac{Dp}{Dt} + p\vec{\nabla}\cdot\vec{u}, \qquad (1.9)$$

and from the enthalpy equation, omitting $\dot{Q}$, we write

$$\rho\frac{Dh}{Dt} = \frac{Dp}{Dt} + \vec{\nabla}\cdot(\lambda\vec{\nabla}T) + 2\mu\left(S_{ij}S_{ij} - \frac{1}{3}S_{ii}S_{jj}\right). \qquad (1.10)$$

It can now easily be shown that by taking the scalar product of the momentum equation (1.6) with $\vec{u}$ and adding the result to Eq. (1.10), we get

$$\rho \frac{D}{Dt}\left(h + \frac{1}{2}\vec{u}^2 + \Phi\right) = \frac{\partial p}{\partial t} + \vec{\nabla} \cdot (\lambda \vec{\nabla} T) + 2\mu\left(S_{ij}S_{ij} - \frac{1}{3}S_{ii}S_{jj}\right)$$

$$+ u_i \frac{\partial}{\partial x_j}\mu\left[\left(\frac{\partial u_i}{\partial x_j} + \frac{\partial u_j}{\partial x_i}\right) - \frac{2}{3}\vec{\nabla} \cdot \vec{u}\, \delta_{ij}\right].$$

(1.11)

Indeed the geopotential Φ is time independent, and thus $D\Phi/Dt = \vec{u} \cdot \vec{\nabla}\Phi$. This gives us the generalized Bernoulli theorem stating that $h + \frac{1}{2}\vec{u}^2 + \Phi$ is an invariant of motion if the flow is time independent and if molecular diffusion is neglected.

For a perfect barotropic fluid (i.e., p is a function of ρ only) where rotation is neglected, the momentum equation reduces to

$$\frac{D\vec{u}}{Dt} = -\vec{\nabla}(h + \Phi).$$

(1.12)

Returning to the more general case, let us consider successively a liquid and a gas.

- For a liquid, we have approximately $e_i = C_p\, T$. However, we can check that the pressure and molecular viscous terms on the right-hand side (r.h.s.) of Eq. (1.8) are in general negligible, and thus we have

$$\frac{DT}{Dt} \approx \kappa \nabla^2 T, \qquad \frac{D\rho}{Dt} \approx \kappa \nabla^2 \rho,$$

(1.13)

  where $\kappa = \lambda/\rho\, C_p$ is the thermal diffusivity.[2]

- For an ideal gas, the state equation reads $p/\rho = RT$ (with $R = C_p - C_v$). We make a further assumption of identifying this thermodynamic pressure with the static pressure already introduced in the stress tensor. We suppose also that C_p and C_v are temperature independent. We now have $e_i = C_v T$. Introducing the potential temperature

$$\Theta = T\left(\frac{p_0}{p}\right)^{(\gamma-1)/\gamma},$$

(1.14)

where $\gamma = C_p/C_v$ and p_0 is the pressure at some reference level, we write

---

[2] Notice, however, that in Eq. (1.13) the density equation is obtained by assuming a linear relation between ρ and T such that ρ is a decreasing function of T. Because of mass conservation this implies that, if T decreases, ρ will increase and the volume of the fluid parcel will decrease. This is no longer true for water at temperatures close to 4 °C, where it will dilate when cooled (Balibar [11]). In this case, pressure effects in Eq. (1.8) have to be taken into account.

the thermodynamic equation as

$$\frac{D\Theta}{Dt} = \frac{\Theta}{C_p T} \left[\frac{1}{\rho} \vec{\nabla} \cdot (\lambda \vec{\nabla} T) + 2\nu \left(S_{ij} S_{ij} - \frac{1}{3} S_{ii} S_{jj} \right) \right]. \qquad (1.15)$$

A good approximation of this equation for subsonic flows is

$$\frac{D\Theta}{Dt} \approx \kappa \frac{\Theta}{T} \nabla^2 T, \qquad (1.16)$$

where the thermal diffusivity κ has the same definition as before for the liquid. We recall that if the motion is adiabatic ($\kappa = 0$), Θ is an invariant of motion, as are both the entropy and $p(\delta V)^\gamma$. If the ideal gas is barotropic (and perfect), it is isentropic.

Validation of these equations of motion comes from the very good comparison of theoretical solutions with laboratory experiments in laminar regimes for cases such as Poiseuille flow in a channel or a pipe, or boundary layers developing over a flat plate, or mixing layers. In the turbulent regimes, first- and second-order statistics of numerical solutions also compare favorably with experiments for the same flows. Only above Mach numbers of the order of 15–20 does the molecular-agitation scale catch up with the continuous-medium scales in such a way that the continuous-medium assumption no longer holds.

The generalized Bernoulli theorem allows us to understand why a hypersonic body heats during atmospheric reentry. Indeed, let us consider a frame fixed to the body and suppose that an upstream fluid parcel is at a velocity U_∞ and a temperature T_∞. Its enthalpy is $C_p T_\infty$. If the parcel hits the body, on which the velocity is zero, neglecting gravity, we get

$$C_p T_\infty + \frac{1}{2} U_\infty^2 = C_p T_a, \qquad (1.17)$$

where T_a is the temperature at the wall, which is higher than T_∞ owing to this exchange between kinetic energy and enthalpy. We will talk more of this adiabatic temperature in the section of Chapter 7 devoted to LES of a space-shuttle rear wing.

Let us finally consider Eq. (1.15) in the case of a compressible, parallel time-independant flow of ideal gas. The velocity-vector components are $[u(y), 0, 0]$. Let $Pr = C_p \mu(y)/\lambda(y)$ be the Prandtl number assumed constant. We have

$$\frac{d}{dy} \left(\mu \frac{dT}{dy} \right) = -\frac{Pr}{C_p} \mu \left(\frac{du}{dy} \right)^2, \qquad (1.18)$$

which shows there is a temperature gradient of molecular-diffusion origin induced by the velocity gradient. This has analogies with the Crocco–Busemann equation. In fact, such a velocity profile is only possible if the pressure $p(x)$

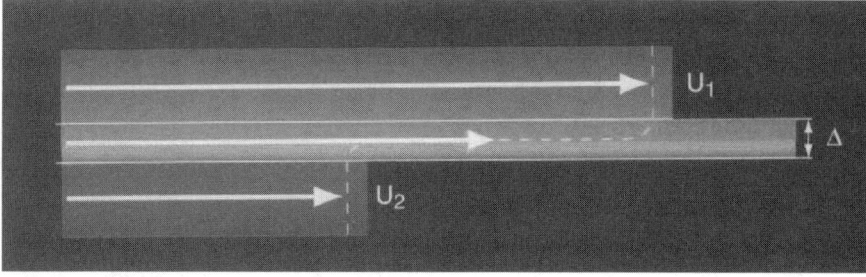

Figure 1.1. Schematic view of a vortex sheet.

depends only on x; then the pressure gradient dp/dx is constant with

$$\frac{dp}{dx} = \frac{d}{dy}\left(\mu\frac{du}{dy}\right).$$

(1.19)

In the weakly compressible case, this yields a parabolic velocity profile if $dp/dx \neq 0$ and a linear velocity profile if $dp/dx = 0$.

1.2.4 Vorticity

A very important quantity for characterizing turbulence (in the absense of entrainment rotation) is the vorticity vector $\vec{\omega} = \vec{\nabla} \times \vec{u}$. A quasi-discontinuity between two parallel flows of velocity $\vec{U}_1$ and $\vec{U}_2$ gives rise to a vortex sheet (see Figure 1.1). The latter is violently unstable under small perturbations (Kelvin–Helmholtz instability) and rolls up into spiral Kelvin–Helmholtz vortices into which vorticity has concentrated. These vortices may undergo secondary successive instabilities, leading to a violent direct kinetic-energy cascade toward small scales; they may also be responsible for inverse energy cascades through pairings (see Lesieur [170], Chapter III). In practive, Kelvin–Helmholtz-type instabilities are the source of turbulence in many hydrodynamic as well as external and internal aerodynamic applications. An illustration is provided by the famous helium–nitrogen mixing-layer experiment carried out at Caltech by Brown and Roshko [33] and presented in Figure 1.2 (top). Figure 1.2 (bottom) shows a "numerical dye" (with the passive scalar of the upstream distribution proportional to the upstream velocity) in a two-dimensional numerical simulation of a uniform-density mixing layer carried out in Grenoble by Normand [220]. We will return in detail to these vortex-dynamic aspects in Chapter 5. Let us focus now on small-scale–developed turbulence characteristics, which are very important to assess the potential of direct numerical simulations of flows in terms of practical applications. This is why we devote a section to very useful spectral tools in isotropic turbulence.

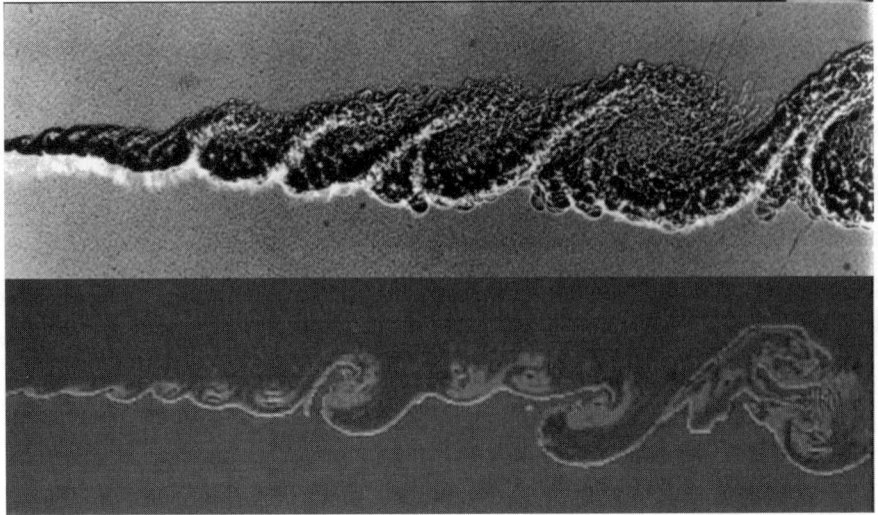

Figure 1.2. (Top) Experimental mixing layer of Brown and Roshko. (Courtesy A. Roshko.) (Bottom) Grenoble two-dimensional numerical simulation. (Courtesy X. Normand.)

1.3 Isotropic turbulence

1.3.1 Formalism

Isotropic turbulence is a model that may be relevant to small-scale–developed turbulent flows. We assume an infinite domain without boundaries. Turbulent quantities are represented by random functions for which averages are taken on ensembles of realizations and are denoted $\langle \ \rangle$. Turbulence is assumed to be statistically invariant under rotations about arbitrary axes (and hence translations). Thus the average velocity is zero. We restrict our attention to a flow of uniform density. The easiest mathematical way to deal with such turbulence is to use spatial Fourier space. Let us first introduce the spatial integral Fourier transform of a given function (scalar or vector) $f(\vec{x}, t)$ associated with turbulence

$$\hat{f}(\vec{k}, t) = \left(\frac{1}{2\pi}\right)^3 \int e^{-i\vec{k}.\vec{x}} \ f(\vec{x}, t) \, d\vec{x}, \qquad (1.20)$$

where the integral is carried out over the entire three-dimensional space. Because turbulence is statistically homogeneous, its fluctuations cannot be expected to decrease at infinity. However, Eq. (1.20) does make sense in the framework of generalized-functions theory (distributions). In this context, the inverse relation

$$f(\vec{x}, t) = \int e^{i\vec{k}.\vec{x}} \ \hat{f}(\vec{k}, t) d\vec{k} \qquad (1.21)$$

also holds. It is interesting to consider the Navier–Stokes equation for a uniform-density flow in Fourier space, which is

$$\left(\frac{\partial}{\partial t} + \nu k^2\right) \hat{u}_i(\vec{k}, t) = -i k_m P_{ij}(\vec{k}) \int_{\vec{p}+\vec{q}=\vec{k}} \hat{u}_j(\vec{p}, t)\hat{u}_m(\vec{q}, t) d\vec{p},$$

(1.22)

where

$$P_{ij}(\vec{k}) = \delta_{ij} - \frac{k_i k_j}{k^2}$$

(1.23)

is the projection tensor on a plane perpendicular to $\vec{k}$, which enables us to eliminate the pressure by respecting the incompressibility constraint. Indeed, incompressibility in Fourier space is written $k_i \hat{u}_i(\vec{k}, t) = 0$.

We define the second-order velocity correlation tensor as

$$U_{ij}(\vec{r}, t) = \langle u_i(\vec{x}, t)u_j(\vec{x} + \vec{r}, t)\rangle.$$

(1.24)

Its Fourier transform is the spectral tensor

$$\hat{U}_{ij}(\vec{k}, t) = \left(\frac{1}{2\pi}\right)^3 \int e^{-i\vec{k}.\vec{r}} \; U_{ij}(\vec{r}, t) \, d\vec{r}.$$

(1.25)

For isotropic turbulence, this can be put in the form

$$\hat{U}_{ij}(\vec{k}, t) = \frac{1}{2} \left[\frac{E(k, t)}{2\pi k^2} P_{ij}(\vec{k}) + i \; \epsilon_{ijs} k_s \frac{H(k, t)}{2\pi k^4}\right],$$

(1.26)

where $E(k, t)$ is the kinetic-energy spectrum and $H(k, t)$ is the helicity spectrum. Let us now introduce some important quadratic quantities of turbulence (per unit mass):

• kinetic energy:

$$E_c(t) = \frac{1}{2}\langle \vec{u}(\vec{x}, t)^2\rangle = \frac{1}{2}\int \hat{U}_{ii}(\vec{k}, t)d\vec{k} = \int_0^{+\infty} E(k, t)dk,$$

(1.27)

• helicity:

$$H_e = \frac{1}{2}\langle \vec{u}(\vec{x}, t).\vec{\omega}(\vec{x}, t)\rangle = \int_0^{+\infty} H(k, t)dk, \quad |H(k, t)| \le kE(k, t),$$

(1.28)

• enstrophy:

$$D(t) = \frac{1}{2}\langle \vec{\omega}^2\rangle = \int_0^{+\infty} k^2 E(k, t)dk.$$

(1.29)

• palinstrophy:

$$P(t) = \frac{1}{2}\langle (\vec{\nabla} \times \vec{\omega})^2\rangle = \int_0^{+\infty} k^4 E(k, t)dk.$$

(1.30)

Enstrophy and energy are related (in the decaying case) by

$$\frac{d}{dt}E_c = -2\nu \; D(t). \tag{1.31}$$

One can derive the following enstrophy evolution equation:

$$\frac{d}{dt}D = \left(\frac{98}{135}\right)^{1/2} s(t) \, D^{3/2} - 2\nu \; P(t), \tag{1.32}$$

where $s(t)$ is defined as minus the skewness of the velocity derivative $\partial u_1/\partial x_1$ in any direction of space:

$$s(t) = -\frac{\left\langle \left(\frac{\partial u_1}{\partial x_1}\right)^3 \right\rangle}{\left\langle \left(\frac{\partial u_1}{\partial x_1}\right)^2 \right\rangle^{3/2}}. \tag{1.33}$$

A constant-skewness model yields enstrophy blowup at a finite time in the Euler case (see [170]).

One can also define the spectrum of a scalar ϑ such that

$$\frac{1}{2}\langle \vartheta^2(\vec{x}, t)\rangle = \int_0^{+\infty} E_\Theta(k, t)dk. \tag{1.34}$$

An important quantity is the kinetic-energy transfer $T(k, t)$ coming from nonlinear interactions in the kinetic-energy spectrum evolution equation

$$\left(\frac{\partial}{\partial t} + 2\nu k^2\right) E(k, t) = T(k, t) + F_o(k), \tag{1.35}$$

where $F_o(k)$ is a possible forcing spectrum, which is a mathematical trick required if one wants to study statistically stationary turbulence. Kinetic-energy conservation by nonlinear terms implies

$$\int_0^{+\infty} T(k, t) = 0. \tag{1.36}$$

The kinetic-energy flux is

$$\Pi(k, t) = \int_k^{+\infty} T(k', t) \, dk' = -\int_0^k T(k', t) \, dk'. \tag{1.37}$$

1.3.2 Kolmogorov $k^{-5/3}$ energy spectrum

Suppose we have a narrow forcing $F_o(k)$ concentrated around a given mode k_i. For a time-independent energy spectrum, Eq. (1.35) can be written as

$$2\nu k^2 \; E(k) = T(k) + F_o(k). \tag{1.38}$$

Let

$$\epsilon = \int_0^{+\infty} F_o(k)dk = 2\nu \int_0^{+\infty} k^2 E(k)dk \qquad (1.39)$$

be the kinetic-energy injection rate equal to the dissipation rate. Integration of Eq. (1.38) from 0 to $k > k_i$ yields

$$\lim_{\nu \to 0} \Pi(k) = \epsilon, \qquad k > k_i, \qquad (1.40)$$

which shows that ϵ is a very important parameter. Kolmogorov assumes then for $k > k_i$ and $\nu \to 0$ that $E(k)$ is a function of ϵ and k only. A dimensional analysis yields

$$E(k) = C_K \epsilon^{2/3} k^{-5/3}, \qquad (1.41)$$

where C_K is called the Kolmogorov constant.

1.3.3 Kolmogorov dissipative scale and wavenumber

The Kolmogorov dissipative scale is obtained in the following way (see [170], Chapter VI, for details). Starting from Kolmogorov's law, we can estimate in physical space a typical velocity difference between two points separated by a distance of r:

$$\delta u(r) \approx (\epsilon \, r)^{1/3}. \qquad (1.42)$$

This is obtained by writing $\epsilon \approx \delta u_r^2/(r/\delta u_r)$. We can thus build a local Reynolds number through the cascade

$$R_r = \frac{r \delta u_r}{\nu} \approx \epsilon^{1/3} r^{4/3} \nu^{-1}. \qquad (1.43)$$

The value of R_r falls below the value of one for

$$r < \eta = \left(\frac{\nu^3}{\epsilon}\right)^{1/4}, \qquad (1.44)$$

which is precisely Kolmogorov's scale, such that motions of smaller wavelength will be damped by viscosity. The Kolmogorov dissipative wavenumber is

$$k_d = \eta^{-1} = \left(\frac{\epsilon}{\nu^3}\right)^{1/4}. \qquad (1.45)$$

Many experiments in the laboratory, the ocean, and the atmosphere show that the kinetic-energy spectrum falls exponentially just before k_d, which is thus a mode above which velocity fluctuations are negligible. This is a strong argument in favor of the absence of singularities in Navier–Stokes equation solutions.

The following dissipative scalings are extremely useful: Let the energy spectrum be written as

$$E(k, t) = \epsilon^{2/3}\eta^{5/3}G(k\eta, t), \tag{1.46}$$

where $G(k\eta, t)$ is a nondimensional function, and $\epsilon^{2/3}\eta^{5/3}$ characterizes a typical dissipative energy-spectrum value. The Kolmogorov-compensated spectrum (which should be a plateau in an exact $k^{-5/3}$ inertial range) is defined by

$$\frac{E(k, t)}{\epsilon^{2/3}k^{-5/3}} = (k\eta)^{5/3}G(k\eta, t) = M(k\eta, t). \tag{1.47}$$

Experimental data show that it has a "Mammoth shape" and that it renormalizes the dissipative range and the high-k part of the inertial range very well.

1.3.4 Integral scale and Taylor microscale

Let $f(r) = \langle u_r(\vec{x}, t)u_r(\vec{x} + \vec{r}, t)\rangle/u'^2$ be the longitudinal velocity correlation coefficient with u_r of rms u' being the velocity component in the $\vec{r}$ direction. The integral scale and associated Reynolds number are defined as

$$l = \int_0^{+\infty} f(r)dr, \quad R_l = \frac{u'l}{\nu}, \tag{1.48}$$

respectively. One finds experimentally that $\epsilon = Au'^3/l$ with $A \approx 1$ (see Tennekes and Lumley [278]). Let us introduce now the Reynolds number $R_\lambda = u'\lambda/\nu$ based on the Taylor microscale λ. The latter characterizes the mean spatial extension of the velocity gradients and is defined by

$$\lambda^2 = \frac{u'^2}{\langle(\partial u_1/\partial x_1)^2\rangle}, \tag{1.49}$$

where u_1 is the velocity fluctuation in any direction (because turbulence is isotropic). We have $\epsilon = \nu\langle(\vec{\nabla} \times \vec{u})^2\rangle = 15\nu\langle(\partial u_1/\partial x_1)^2\rangle = Au'^3/l$, and Eq. (1.49) yields

$$\lambda = \sqrt{\frac{15\nu l}{Au'}}, \quad R_\lambda = \sqrt{\frac{15}{A}}\sqrt{\frac{u'l}{\nu}} \, ;$$

hence,

$$R_\lambda = \sqrt{\frac{15}{A}}\sqrt{R_l} \sim \frac{l}{\lambda}. \tag{1.50}$$

These scales and Reynolds numbers are important for evaluating the cost of direct numerical simulations of turbulence. Indeed, a scale characteristic of the

smallest scales is the Kolmogorov dissipative scale η. The number of degrees
of freedom of turbulence in each direction of space may be calculated as

$$\frac{k_d}{k_i} \approx lk_d = \frac{l}{\eta} = l\left(\frac{\epsilon}{\nu^3}\right)^{1/4} = A^{1/4} R_l^{3/4}, \qquad (1.51)$$

which yields for the entire three-dimensional space

$$N_d \approx R_l^{9/4} \sim R_\lambda^{9/2}. \qquad (1.52)$$

1.4 Direct numerical simulations of turbulence

In 1922, the meteorologist Richardson [243] proposed numerical schemes to
solve in a deterministic fashion the equations of fluid mechanics applied to
the atmosphere. This marked the beginning of direct numerical simulations
of turbulence, which are deterministic time-advancing numerical solutions of
fluid mechanics equations with a proper set of initial and boundary conditions.
This is possible[3] provided the two following conditions are fulfilled:

* the numerical schemes are accurate enough, and
* all the scales of motion, from the largest to the smallest, are captured.

There is some evidence that small-scale turbulence is not far from isotropy
even if large scales are not (see Jimenez [136]), and thus λ may be evaluated
even for nonisotropic flows: Jimenez stresses that $R_\lambda \approx 3,000$ in the boundary layer of a commercial aircraft, 10^4 in the atmospheric boundary layer,
and higher values are present in astrophysics. Using Eq. (1.52), we find that this
entails, respectively, using more than 10^{15} and 10^{18} points in computer simulations for the two cases. At present, to avoid excessive computing times on even
the biggest machines, one has to restrict calculations to about 2×10^7 grid
points, which are many orders of magnitude shy of these estimates. Even
with the unprecedented improvement of scientific computers, it may take several decades (if it ever becomes possible) before DNS permits us to capture
situations at Reynolds numbers comparable to those encountered in natural
conditions. This demonstrates the immense interest in LES techniques.

1.5 A brief history of LES

The history of LES began in the 1960s with the introduction of the famous
Smagorinsky's eddy viscosity [269] proposed in 1963. Smagorinsky, a meteorologist like Richardson, did work in the famous mathematical-modeling
group founded by von Neuman. In fact, Smagorinsky wanted to represent the

---

[3] However, one must assume a "Laplacian" point of view of existence and uniqueness of
  solutions at arbitrary times, which is not proven mathematically in three dimensions.

effects on a quasi-two-dimensional, large-scale atmospheric or oceanic flow of a three-dimensional subgrid-scale turbulence following a Kolmogorov direct cascade. It is interesting to remark that Smagorinsky's model was a total failure as far as atmospheric and oceanic dynamics are concerned because it overly dissipates the large scales. Therefore large-scale atmospheric or oceanic numerical modelers turned toward hyperviscous subgrid models. Nonetheless, Smagorinsky's model was extensively used by people interested in industrial applications (and also small- or mesoscale meteorology), which shows that the outcome of research may be as unpredictable as turbulence itself. One should mention the important contribution of Lilly, another meteorologist and Smagorinsky's collaborator, who calculated the value of the Smagorinsky constant in terms of the Kolmogorov constant in three-dimensional isotropic-developed turbulence [184]. In fact, in 1962 Lilly [182] published a LES of buoyant convection in the atmosphere using Smagorinsky's model. The first application of the latter to engineering flows was the pioneering study of a plane channel done by Deardorff [63], another meteorologist, who with his collaborators started at the same time an impressive series of works on large-eddy simulations of the planetary boundary layer (Deardorff [64], Somméria [271]). The last three works use transport equations for the subgrid quantities derived from a one-point closure-type analysis and are in fact precursors of the so-called unstationary Reynolds-averaged Navier–Stokes equations that have been developed these past few years (see the book by Durbin Chichester and Pettersson Reif Chichester [83] devoted to these methods and the discussion later). In 1975 Schumann [259] applied the SGS kinetic-energy transport equation to LES of plane channels and annuli together with a method to account for the anisotropy of the finite-difference grid. He found in the case of the turbulent plane channel a very good agreement with the experiments of Laufer [158] and Comte-Bellot [49] as far as the mean and rms velocities are concerned.

Meanwhile, the theoretical physicist Kraichnan[4] developed in 1976 the important concept of spectral eddy viscosity [147]. This was done in the context of isotropic turbulence two-point closures, and it allowed the calculation of the kinetic-energy transfers between a given Fourier mode $k < k_m$ and modes greater than k_m. In the framework of LES, and as already noticed by Lilly, who used this fact to determine the Smagorinsky constant as a function of the Kolmogorov constant, subgrid-scale wavelengths smaller than Δx correspond in Fourier space to high spatial frequencies larger than a cutoff mode k_C of the order of $\pi / \Delta x$.

The spectral eddy viscosity, utilized now at the level of the Navier–Stokes equation (and no longer for kinetic-energy spectrum studies), was used in

---

[4] Kraichnan worked as a postdoctoral student with Einstein at Princeton.

Grenoble by Lesieur and co-workers to develop LES in spectral space. Such an eddy viscosity has the great advantage of overcoming the scale-separation assumption inherent to any eddy-viscosity model in physical space. They carried out the first LES of decaying three-dimensional isotropic turbulence (Chollet and Lesieur [41]). They also extended the notion of spectral eddy viscosity to a spectral eddy diffusivity and did the first LES of passive-scalar decay in isotropic turbulence [42]. We stress that LES of scalars (passive or active) is essential for mixing and combustion studies. These spectral eddy-viscosity and diffusivity models were adapted to physical space in the form of the structure-function model (Métais and Lesieur [205]), with better results for three-dimensional isotropic turbulence than those of Smagorinsky in terms of Kolmogorov $k^{-5/3}$ inertial range. For applications to shear flows (free or wall-bounded), the structure-function model was adapted to filter out the inhomogeneous effects of the larger scales: This yielded, respectively, the selective [61] and filtered [81] structure-function models. Meanwhile, the physical-space eddy-viscosity concept was revived in a joint work between Stanford and Torino, with a dynamic evaluation of Smagorinsky's constant through a double filtering (Germano and co-workers [108], [109]). The dynamic procedure was also associated with the scale-similarity ideas of Bardina et al. [12] to extend the eddy-viscosity concept.

1.6 LES and determinism

Turbulence in fluids is still considered one of the most difficult problems posed in physics. Let us recall in particular Feynman's statement that "turbulence [is] the last great unsolved problem of classical physics." One is, however, far from the complexity of molecular microscopic physics, since one just deals with the laws of Newtonian mechanics applied to a continuous medium in which molecular-diffusion effects have been filtered out and replaced by molecular viscous exchanges. Such a system has a double behavior of determinism in the Laplacian sense and extreme sensitivity to initial conditions because of its strong nonlinearity.

From a mathematical viewpoint, the LES problem is not very well posed. Indeed, let us consider the time evolution of the fluid as the motion of a point in a sort of phase space of extremely large dimension (e.g., $\sim 10^{15}$ around a wing). Suppose that at some initial instant t_0, the LES flow is taken to be identical to the exact flow in the resolved scales. In scales smaller than Δx, the LES motion is not defined. As stressed by Lesieur ([170], p. 380), let us consider two realizations of the actual flow, identical to the LES in the large scales and completely decorrelated in the subgrid scales. If we accept the results on the propagation of unpredictability caused by nonlinear effects, the difference between the two fields will propagate into the large scales by error backscatter, and the two exact realizations will depart in these scales. Now let us assume that we have been able to solve the subgrid-scale modeling problem posed by LES and dispose of closed large-scale equations in which everything is expressed in terms of these scales. Then the LES field will evolve as a third realization in phase space different from the two other actual realizations in resolved scales. So, as time goes on, the LES will depart from reality. However, as will be seen in the following, LES enables us to predict the statistical characteristics of turbulence as well as the dynamics of coherent vortices and structures.

Note that chaos in dynamical systems with a low number of degrees of freedom is generally characterized by a positive Lyapounov exponent with exponential growth of the distance between two points initially very close in phase space. In isotropic turbulence, we introduce for predictability studies the error spectrum $E_\Delta(k, t)$, characterizing the spatial-frequency distribution associated with the energy of the difference between two random fields $\vec{u}_1$ and $\vec{u}_2$ with same statistical properties,

$$\frac{1}{4}\langle[\vec{u}_1^2(\vec{x}, t) - \vec{u}_2^2(\vec{x}, t)]\rangle = \int_0^{+\infty} E_\Delta(k, t)\, dk, \qquad (1.53)$$

where the energy spectrum $E(k, t)$ is such that

$$\frac{1}{2}\langle\vec{u}_1^2\rangle = \frac{1}{2}\langle\vec{u}_2^2\rangle = \int_0^{+\infty} E(k, t)\, dk. \qquad (1.54)$$

The error rate

$$r(t) = \frac{\int_0^{+\infty} E_\Delta(k, t)dk}{\int_0^{+\infty} E(k, t)dk} \qquad (1.55)$$

equals zero when the two fields are completely correlated and equals one when they are totally uncorrelated. In predictability studies, one generally takes an initial state such that complete unpredictability [$E(k) = E_\Delta(k)$] holds above $k_E(0)$, while $E_\Delta(k)$ is 0 for $k < k_E(0)$. Two-point closures of the EDQNM type (see [170] for details) show (in three or two dimensions) an inverse cascade of error, where the wavenumber $k_E(t)$ characterizing the error front decreases (see Métais and Lesieur [204]). Thus, the error rate can be approximated by

$$r(t) \approx \frac{\int_{k_E(t)}^{\infty} E(k, t)dk}{\int_0^{+\infty} E(k, t)dk}. \qquad (1.56)$$

We assume that the turbulence is forced by external forces; therefore, the kinetic energy arising in the denominator of Eq. (1.56) is fixed. In three-dimensional turbulence, and if a $k^{-5/3}$ spectrum is assumed for $k > k_E$, the error rate will be proportional to $\int_{k_E}^{\infty} k^{-5/3}dk \sim k_E^{-2/3}$. In fact, closures show that k_E^{-1} follows a Richardson's law ($k_E^{-1} \propto t^{3/2}$), and thus the error rate grows linearly with time. A similar behavior has been reported by Hunt [130] in an analysis of weather forecast models. This is in fact a slow increase compared with the exponential growth of chaotic dynamical systems and is quite encouraging from the standpoint of potentially using LES for three-dimensional turbulent flows.

1.7 The place of LES in turbulence modeling

Let us call turbulence numerical modeling any numerical approach allowing us to predict the evolution of instantaneous or mean quantities associated with turbulence and that can be applied to shear flows.

The first approach is DNS, which provides both instantaneous and statistical predictions. It is exact[5] but very costly and is limited to low Reynolds number flows, as previously discussed.

The second approach is LES. It is a type of DNS of the large scales, and it gives, as does DNS, both instantaneous and statistical data.

The third approach consists of solving the Reynolds equation, an ensemble-averaged Navier–Stokes equation for which the ensemble average is the same as previously introduced for isotropic turbulence and is taken on an infinite ensemble of independant realizations. The Reynolds equation (at least in the constant-density case) is the Navier–Stokes equation for $\langle u_i \rangle$ with extra stresses given by the Reynolds stresses. This is what people call RANS (Reynolds-averaged Navier–Stokes). The very difficult one-point closure problem previously mentioned consists of modeling the Reynolds stresses in terms of $\langle u_i \rangle$. In the case of all statistically stationary turbulent flows (such as a wake, mixing layer, backstep, or jet), the relevant solutions of the Reynolds equation cannot have by essence any time dependance because the flow is statistically stationary. However, numerical solutions of the Reynolds equations closed by some model and with the velocity time derivative retained show unstationary phenomena such as shedding of Karman vortices in a wake or Kelvin–Helmholtz vortices in a mixing layer or a backstep. In fact it can be shown that, when a clear shedding frequency of vortices may be identified in the flow, phase averaging of the instantaneous velocity with respect to the period of the shedding gives rise to an unstationary equation similar to the Reynolds equation. Such a formalism was introduced by Reynolds and Hussain [242]. This may justify the unstationary solutions found for the modeled Reynolds equation. However, phase averaging is not a well-defined operator downstream of shear flows, where coherent vortices become unpredictable. It cannot be defined at all in isotropic turbulence if no solution of the flow is known. The problem is that wide use is currently made of the

---

[5] The exactness holds only if numerical schemes are accurate enough.

so-called unstationary RANS methods for complex industrial or environmental turbulent flows without due consideration of the significance of the computed solutions. Considering the strong analogy of LES equations and unstationary RANS equations at the level of linear momentum, we think that the latter approach should be viewed more as a loosely resolved LES. Comparisons of both methods with the same numerical code and the same resolution should be carried out for a wide range of flows to evaluate the role of transport equations in RANS. We stress again that LES of an unstationary RANS type has already been used for a long time by Deardorff [65] and co-workers for meteorological applications as well as by Schmidt and Schumann [258]. The latter studied the structure of turbulent thermal convection in the atmosphere. A problem of the same type was addressed by Hanjalic and Kenjeres [117] using unstationary RANS. They extended their work in the magnetohydrodynamic case by adding a magnetic field, with very impressive visualizations showing the influence of the Hartmann number on the flow structure.

2 Vortex dynamics

As was already briefly discussed, large-eddy simulations deal with energetic structures of the flow with a characteristic scale or wavelength larger than a given cutoff scale Δx. These so-called large scales may or may not be spatially organized and sometimes correspond to coherent vortices of recognizable shape. It is therefore important that we use precise language. Within these large scales, we will consider in particular *coherent vortices* and *coherent structures*. However, some of the large scales do not fall into these two categories. In this respect, the term "large-eddy simulations" is not very well chosen, and the French expression "simulation des grandes échelles" (large-scale simulations) is more appropriate.

Before looking more specifically at coherent vortices, it is of interest to recall the basic elements of vorticity dynamics associated with the behavior of the vorticity vector $\vec{\omega}$.

2.1 Vorticity dynamics

Taking the curl of the momentum equation (1.7), we have

$$\frac{\partial \vec{\omega}_a}{\partial t} + \vec{\nabla} \times (\vec{\omega}_a \times \vec{u}) = \frac{1}{\rho^2} \vec{\nabla}\rho \times \vec{\nabla}p + \text{viscous dissipation}, \qquad (2.1)$$

which may be written as

$$\frac{D\vec{\omega}_a}{Dt} = \vec{\omega}_a \cdot \vec{\nabla}\vec{u} - (\vec{\nabla} \cdot \vec{u})\,\vec{\omega}_a + \frac{1}{\rho^2} \vec{\nabla}\rho \times \vec{\nabla}p + \text{viscous dissipation},$$
$$(2.2)$$

or, using the continuity equation, as

$$\frac{D}{Dt}\left(\frac{\vec{\omega}_a}{\rho}\right) = \frac{\vec{\omega}_a}{\rho} \cdot \vec{\nabla}\vec{u} + \frac{1}{\rho^3} \vec{\nabla}\rho \times \vec{\nabla}p + \text{viscous dissipation}, \qquad (2.3)$$

where

$$\frac{\vec{\omega}_a}{\rho} \cdot \vec{\nabla}\vec{u} = \vec{\nabla}\vec{u} \otimes \left(\frac{\vec{\omega}_a}{\rho}\right).$$

Hence, for a perfect barotropic fluid, $\vec{\omega}_a/\rho$ satisfies the equation of evolution of a small vector $\overrightarrow{MM'}$ when M and M' follow the fluid motion. Indeed

$$\frac{D}{Dt}\overrightarrow{MM'} = \vec{u}(M') - \vec{u}(M) = \overrightarrow{MM'} \cdot \vec{\nabla}\vec{u} = \vec{\nabla}\vec{u} \otimes \overrightarrow{MM'} \qquad (2.4)$$

characterizes the "passive-vector" equation valid to the first order in $\overrightarrow{MM'}$.

2.1.1 Helmholtz–Kelvin's theorem

Let us consider the circulation of the velocity along a closed contour C. We have

$$\frac{D}{Dt}\oint_C \vec{u} \cdot \vec{\delta l} = \oint_C \vec{u} \cdot \frac{D\vec{\delta l}}{Dt} + \oint_C \frac{D\vec{u}}{Dt} \cdot \vec{\delta l}. \qquad (2.5)$$

The first term on the right-hand side (r.h.s.) of Eq. (2.5) is zero because of Eq. (2.4), which applies to $\vec{\delta l}$. The second will be zero if $D\vec{u}/Dt$ is proportional to a gradient. This happens in two cases (for perfect flow, and no rotation): if ρ is uniform (Helmholtz's theorem), or if the flow is barotropic (Kelvin's theorem). This implies that, in the conditions of the theorem, vortex tubes (whose envelope comprises vortex lines tangent to the vorticity vector at each point) are material and travel with the fluid parcels they contain.

2.2 Coherent vortices

For the rest of the chapter, we will discuss nonrotating flows ($\vec{\Omega} = 0$) of uniform density ρ_0.

2.2.1 Definition

Coherent vortices in turbulence are defined by Lesieur ([170], pp. 6–7) as regions of the flow satisfying three conditions:

(i) The concentration of ω, modulus of the vorticity vector, should be high enough so that a local rollup of the surrounding fluid is possible.
(ii) They should keep their shape approximately during a time T_c long enough in front of the local turnover time ω^{-1}.
(iii) They should be unpredictable.

   In this context, high ω is a possible candidate for coherent-vortex identification.

2.2.2 Pressure

With such a definition, the cores of the coherent vortices should be pressure lows. Indeed, a fluid parcel winding around the vortex will be (in a frame moving with the parcel) in approximate balance between centrifugal and pressure-gradient effects. We are talking here of the static pressure p. The reasoning may be made more quantitative by considering the Euler equation (in a flow of uniform density ρ_0) in the form

$$\frac{\partial \vec{u}}{\partial t} + \vec{\omega} \times \vec{u} = -\frac{1}{\rho_0} \vec{\nabla} P, \qquad (2.6)$$

where $P = p + \rho_0 [\Phi + (\vec{u}^2/2)]$ is now the dynamic pressure. In a frame moving with the coherent vortex and supposed locally Galilean, the ratio (within the vortex) of the second to the first term on the left-hand side (l.h.s.) of Eq. (2.6) is of the order of $T_c\,\omega$. Then the equation reduces for the coherent vortex to the cyclostrophic balance

$$\vec{\omega} \times \vec{u} \approx -\frac{1}{\rho_0} \vec{\nabla} P \qquad (2.7)$$

if condition (ii) is fulfilled. If one supposes that the coherent vortex is a vortex tube tangent to the velocity vector, it follows that this tube is a low for the dynamic pressure.

2.2.3 The Q-criterion

We recall now the so-called Q-criterion. Let

$$S_{ij} = \frac{1}{2}\left(\frac{\partial u_i}{\partial x_j} + \frac{\partial u_j}{\partial x_i}\right), \quad \Omega_{ij} = \frac{1}{2}\left(\frac{\partial u_i}{\partial x_j} - \frac{\partial u_j}{\partial x_i}\right) \qquad (2.8)$$

be, respectively, the symmetric and antisymmetric parts of the velocity-gradient tensor $\partial u_i/\partial x_j$. It is well known that the second invariant of this tensor,

$$Q = \frac{1}{2}(\Omega_{ij}\Omega_{ij} - S_{ij}S_{ij}) = \frac{1}{4}(\vec{\omega}^2 - 2S_{ij}S_{ij}), \qquad (2.9)$$

is equal to $\nabla^2 p/2\rho_0$. Indeed, the Poisson equation for the pressure in a flow of uniform density can be written as

$$-\frac{\nabla^2 p}{\rho_0} = \frac{\partial^2}{\partial x_i \partial x_j} u_i u_j = \frac{\partial}{\partial x_i}\left[u_j \frac{\partial u_i}{\partial x_j}\right] = \frac{\partial u_i}{\partial x_j}\frac{\partial u_j}{\partial x_i}$$

$$= \left(S_{ij} + \frac{1}{2}\epsilon_{ij\lambda}\omega_\lambda\right)\left(S_{ji} + \frac{1}{2}\epsilon_{ji\mu}\omega_\mu\right) = S_{ij}S_{ij} - \frac{1}{2}\vec{\omega}^2 = -2Q.$$

Let us present now a line of reasoning discussed in [78]. We consider a low static-pressure tube of small section (see Figure 2.1). Let $\Delta\Sigma$ be its lateral

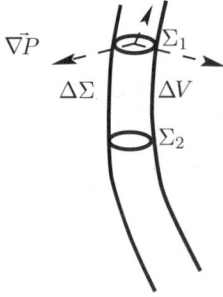

Figure 2.1. Schematic low-pressure tube. (From Dubief and Delcayre [78]; courtesy *Journal of Turbulence*.)

surface, which is assumed to be isobaric and convex. Let Σ_1 and Σ_2 be two cross sections of the tube normal to its axis, and let ΔV be the volume of the tube portion between Σ_1 and Σ_2. The pressure gradient on $\Delta \Sigma$ is normal to it and directed toward the exterior. The pressure gradient on the two cross sections is tangent to them. Then, the flux of the pressure gradient getting out of the tube is equal to the flux through $\Delta \Sigma$ and is positive. From the divergence theorem, this is equal to the integral over ΔV of $\nabla^2 p$, which is positive, as well as the integral of Q. If we suppose that the size of ΔV is small enough that Q does not vary appreciably within it, this implies that Q is positive in ΔV. This reasoning may be repeated all along the length of the tube, and the Q-criterion ($Q > 0$) is therefore a necessary condition for the existence of such thin, convex low-pressure tubes.

To our knowledge, the Q-criterion was first proposed by Weiss [290] to characterize "elliptic" regions in two-dimensional turbulence. Let us consider the inviscid vorticity-gradient equation in this case,

$$\frac{D}{Dt}\vec{\nabla}\omega = -\vec{\nabla}\vec{u}|^{\mathrm{t}} \otimes \vec{\nabla}\omega, \qquad (2.10)$$

where here ω is the vertical vorticity component and $\vec{\nabla}\vec{u}|^{\mathrm{t}}$ is the transposed velocity-gradient tensor. We can calculate the eigenvalues of $\vec{\nabla}\vec{u}|^{\mathrm{t}}$, which are identical to those of $\vec{\nabla}\vec{u}$. Their square is $-Q$, and they are purely imaginary if $Q > 0$. In this case the vorticity gradient will rotate locally – a property that is expected from a vortex. It was checked by Basdevant and Philipovitch [16] using DNS that the core of vortices in two-dimensional isotropic turbulence is quite well represented by Weiss's elliptic regions.[1] We notice that Eq. (2.10) is also valid for a passive-scalar gradient $\vec{\nabla}\rho$ if one neglects molecular diffusion,

$$\frac{D}{Dt}\vec{\nabla}\rho = -\vec{\nabla}\vec{u}|^{\mathrm{t}} \otimes \vec{\nabla}\rho, \qquad (2.11)$$

and the scalar gradient will also rotate locally in elliptic regions.

In three dimensions, Eq. (2.10) is no longer valid of course, but Eq. (2.11) is. We recall the very important work of Chong et al. [44], which is well

---

[1] The quantity Weiss called Q was in fact of opposite sign with respect to the present notation.

reviewed by Bernard and Wallace ([22], p. 167). We write the passive-vector equation (2.4) for a vector $\vec{\delta l}$ lying in a plane perpendicular to the vortex axis, whose origin is located on the latter and whose extremity is slightly away:

$$\frac{D}{Dt}\vec{\delta l} = \vec{\nabla} \vec{u} \otimes \vec{\delta l}. \tag{2.12}$$

Within a vortex, the extremity of $\vec{\delta l}$ should rotate. Chong et al. [44] have shown that this occurs when the tensor $\vec{\nabla} \vec{u}$ has one real eigenvalue and two complex ones. In fact, the eigenvalues are, in the constant-density case, the solution of the equation

$$\lambda^3 + Q\lambda + R = 0, \tag{2.13}$$

where R is minus the determinant of the matrix associated with the velocity-gradient tensor. The conditions sought for the eigenvalues correspond to a positive discriminant

$$4Q^3 + 27R^2 > 0. \tag{2.14}$$

This so-called Δ-criterion is complicated, and the simplified approximate condition $Q > 0$ was proposed in Hunt et al. [129] to characterize the vortices. Notice that in two dimensions we may apply the same analysis for the vector $\vec{\delta l}$. Equation (2.13) reduces to $\lambda^2 + Q = 0$. We then recover Weiss's result.

   We have thus shown that the Q-criterion, although not exact in three dimensions for characterizing the local rotation of a passive-scalar gradient, is valuable to help characterize convex low-pressure tubes, which are generally associated with coherent vortices. Notice, however, that the relation $Q = \nabla^2 p/2\rho_0$ implies that vortex-identification criteria based on Q involve much more small-scale activity than those based on the pressure, as will be verified in the simulations. It is also clear from Eq. (2.14) that positive Q implies that the Δ-criterion is fulfilled. The latter is hence more restrictive than the Q-criterion.

2.2.4 The λ_2-criterion

We now briefly describe the λ_2-criterion introduced by Jeong and Hussain [133]. They consider the evolution equation for S_{ij}, which in the Euler case (see, e.g., Ohkitani [222]) can be written as

$$\frac{D}{Dt}S_{ij} + \Omega_{ik}\Omega_{kj} + S_{ik}S_{kj} = -\frac{1}{\rho_0}p_{,ij}, \tag{2.15}$$

where the notation $, i$ stands for a derivative with respect to x_i. The quantity $p_{,ij} = \partial^2 p / \partial x_i \partial x_j$ is called the pressure Hessian. Jeong and Hussain neglect the Lagrangian time-derivative term[2] and associate a coherent vortex to a local minimum of pressure. In fact the pressure Hessian is a real symmetric matrix whose eigenvalues are real, and the fact that pressure is minimum implies that two of these eigenvalues are positive. This implies that $\Omega_{ik}\Omega_{kj} + S_{ik}S_{kj}$ (which is also a real symmetric matrix and has real eigenvalues) has two negative eigenvalues and hence that its second eigenvalue λ_2 is negative. Because we have shown that Q is positive within small low-pressure tubes of convex cross section, the Q and λ_2 criteria can be said to be strongly related (see also [59]).

2.2.5 Simple two-dimensional vortex interactions

Let us briefly recall two essential two-dimensional vortex interactions, which will turn out to be important even in three-dimensional shear flows (free or wall-bounded): pairings of same-sign vortices and traveling dipoles of opposite-sign vortices.

Let us first consider in a plane a local vorticity concentration within a convex closed contour C. Let Σ be the enclosed area, and let $\bar{\omega}\Sigma$ be the vorticity flux across the surface. The fluid outside of C is assumed to be irrotational. Stokes circulation theorem implies the existence of an induced azimuthal velocity within the irrotational region, whose modulus a distance r apart from the vortex is $\approx \bar{\omega}\Sigma/2\pi r$.

- Pairing: Suppose we have two same-sign vortices. They will revole around each other owing to velocity induction, and they will pair if they are close enough. During the process, they form spiral arms reminiscent of galaxies owing to the differential rotation[3] existing between the interior and the exterior of the vortices. Many pairings are observed in mixing-layer experiments or computations such as those shown in Chapter 1.
- Dipoles: Opposite-sign vortices will travel together because of the mutual induced velocity. This phenomenon enables us to explain the self-raising of traveling hairpin vortices (see the present chapter and Chapter 7).

2.3 Vortex identification

Let us present a comparison of some of these vortex-identification methods (low pressure, high $|\omega|$, positive Q, and negative λ_2) applied to incompressible DNS of isotropic turbulence and LES of a backward-facing step as done by Delcayre [69]. Other exemples will be provided in the rest of the book. More specifically, we consider isosurfaces at a given threshold of $|\omega|$, P, Q, and

---

[2] This is acceptable if the vortex is coherent enough.
[3] The angular velocities are $\propto r^{-2}$.

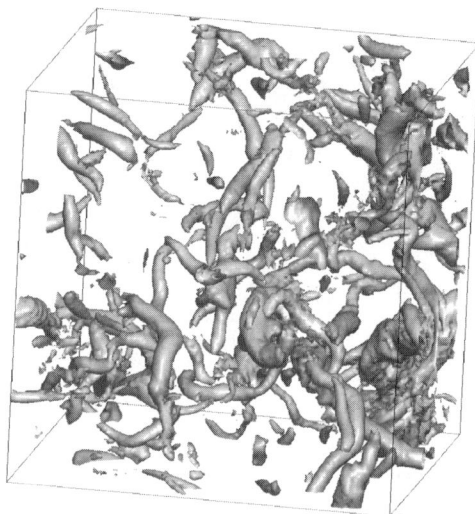

Figure 2.2. Low-pressure isosurfaces in DNS of isotropic turbulence. (From Delcayre [69].)

λ_2. The choice of the threshold is justified by what visually gives the best vortices, or with respect to what we know of the flow dynamics from former simulations or laboratory experiments.

2.3.1 Isotropic turbulence

For isotropic turbulence, we first consider a DNS at low Reynolds number (freely decaying case) done by Lesieur et al. [171] using pseudo-spectral methods. We start from a Gaussian initial velocity, whose coherent vortices during the self-similar decay period following the enstrophy blowup have been analyzed by Delcayre [69]. It is well known that coherent vortices exist in such a flow in the form of randomly oriented thin tubes of length equal to the turbulence integral scale (see, e.g., Siggia [263], She et al. [262], Vincent and Ménéguzzi [288], Métais and Lesieur [205], Jimenez and Wray [135]). Comparison of Figures 2.2 and 2.3 (left) shows that the isobaric surfaces are fatter than the vorticity surfaces but represent the same large-scale events, which are findings in good agreement with the observations of Brachet [29] for Taylor–Green vortices and Métais and Lesieur [205] for LES of isotropic turbulence. Let us mention also the laboratory experiments of Cadot et al. [35] involving turbulence between two counterrotating disks, which displayed the presence of vortices that were pressure troughs. Figure 2.3 (right), showing the iso-Q maps, is close to the vorticity map, although it is slightly less dense. The negative λ_2 map (not presented here) resembles the Q maps.

The first three movies on the CD-ROM present animations of, respectively, low pressure (Animation 2-1), positive Q (Animation 2-2), and vorticity norm (Animation 2-3) in a LES of decaying isotropic turbulence using the spectral-dynamic model (see Chapter 4) carried out in [176] with

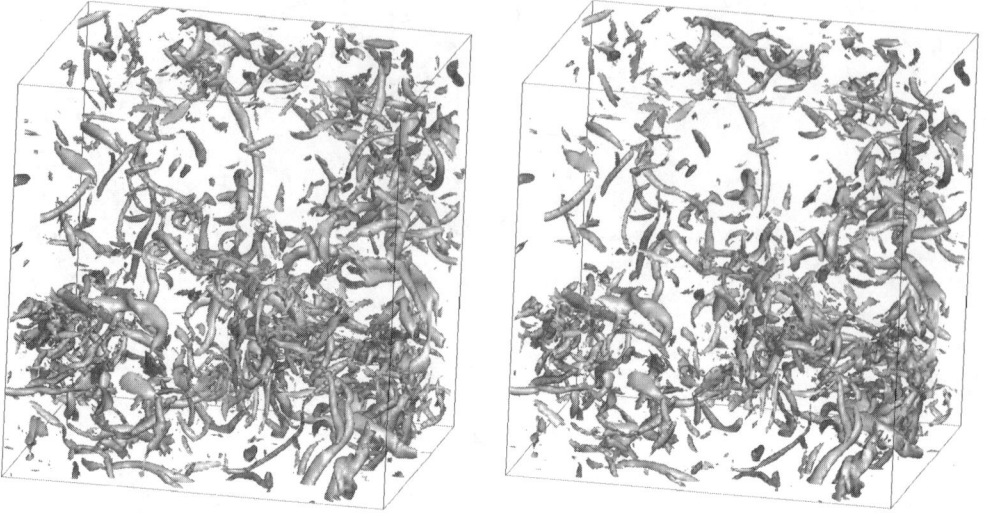

Figure 2.3. High-vorticity (left) and positive Q (right) isosurfaces in DNS of isotropic turbulence. (From [69].)

128^3 collocation points. The initial peak of the kinetic-energy spectrum is at $k_i = 4$. The evolution goes from $t = 0$ to 15 initial large-eddy turnover times (see Chapter 4 for more details on isotropic turbulence evolution). Here again, the initial velocity field is Gaussian. The threshold values for P, Q, and the vorticity are chosen empirically to give the best visual representation of vortices. The pressure animation (see also Figures 2.4 and 2.5) starts with a few big low-pressure structures in the form of billows and even bubbles, some of which seem to be attached to the billows. These structures are associated with the initial nondivergent Gaussian field. These big Gaussian structures evolve and interact in a complicated and difficult-to-follow manner in such a way as to become thinner and thinner. At $t = 7$ they have nearly totally disappeared – at least as far as the particular threshold is concerned.[4] In Figures 2.6–2.8, which are fixed views at $t = 6$ taken from Animation 2-1, the lower and left sides of the computational box are colored by the value of the associated quantity on this side. Figure 2.6 displays the pressure.[5] One notices on the left side of the box an initial low-pressure peak (due to initial conditions) whose intensity diminishes, then grows again at about $t = 2$, and then decreases. In the Q evolution of the animation and of Figures 2.4 and 2.5, nothing is seen at the initial instant. Then one sees the progressive formation of tubes (much thinner than the pressure tubes), which have filled the space at $t = 4$. Beyond this time, one sees the rapid appearance of small-scale turbulence, which seems to be due to

---

[4] See the discussion on time-varying thresholds that follows.
[5] It is in fact a macropressure, whose definition will be given in the next chapter.

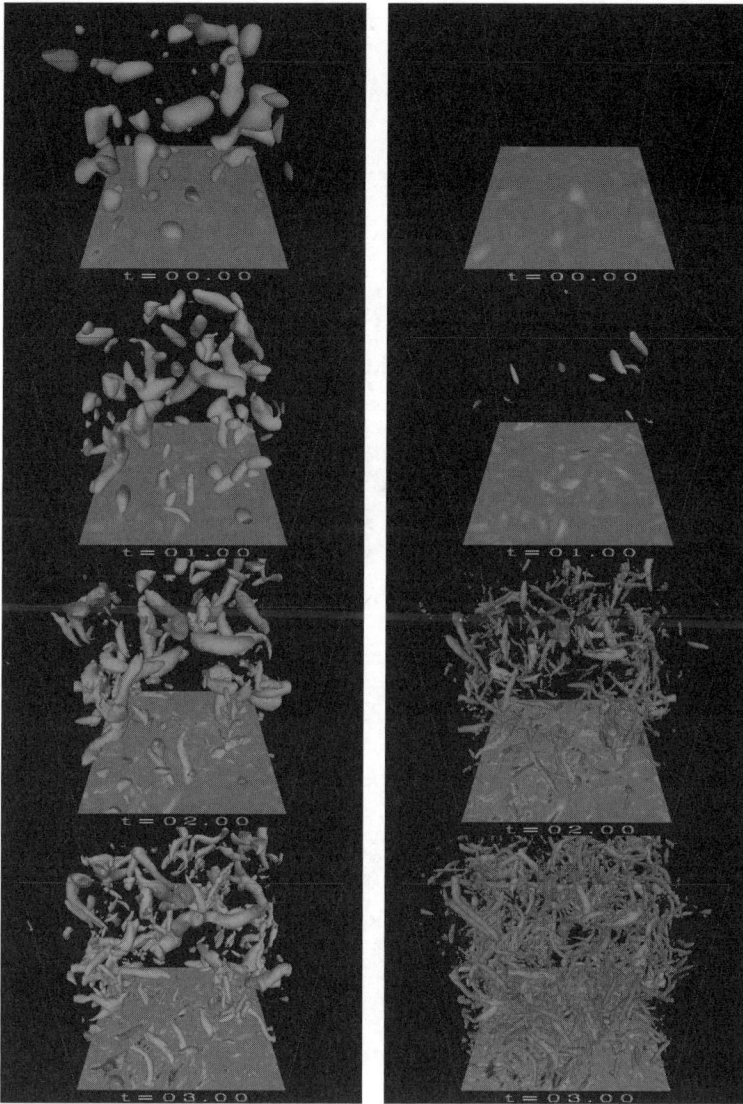

Figure 2.4. Decaying isotropic turbulence. Successive evolution of pressure (left) and Q (right) from $t = 0$ to $t = 3$.

the breakdown of larger scale tubes in some regions of the flow and is finished at $t = 5$. Afterward, one observes a superposition of large-scale and fine-scale tubes, as well as other small scales not organized into tubes. Turbulence seems to be more intermittent in the sense that coherent structures occupy a smaller fraction of space. The animation of vorticity (see also Figure 2.8) shows in the same conditions the evolution of the vorticity modulus. One sees hardly any difference when comparing with Q, and the formation of vortex sheets, which by rollup would generate the coherent vortices, is not obvious. On the

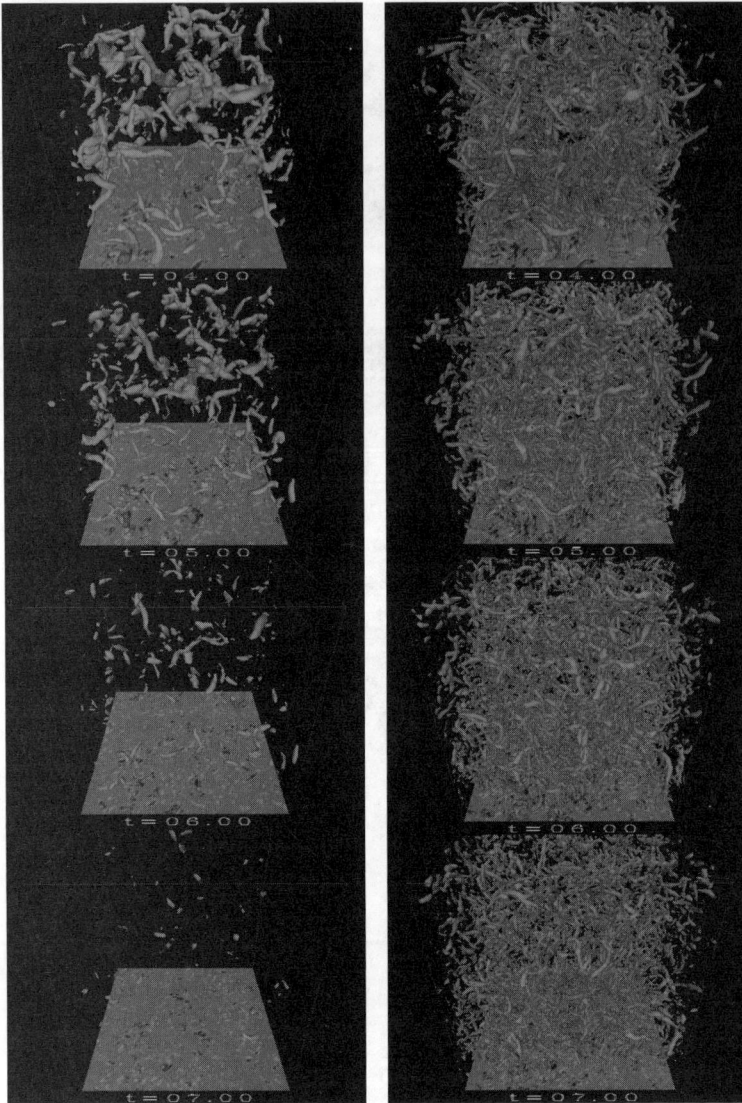

Figure 2.5. Decaying isotropic turbulence. Successive evolution of pressure (left) and Q (right) from $t = 4$ to $t = 7$.

left side of the box, and in contrast to the amplitude of the pressure troughs, the intensity of high vorticity increases continuously during several turnover times. It is therefore clear in this case that there is no correlation between low pressure and high vorticity during the initial stage of evolution of such turbulence initially close to Gaussianity.

Let us return to Figures 2.4 and 2.5, which show in fact the "birth and evolution" of vortices through pressure and Q. It should be stressed that for large times and if a time-varying threshold adjusting to the decaying vortex intensity was chosen, one should certainly be able to observe vortices.

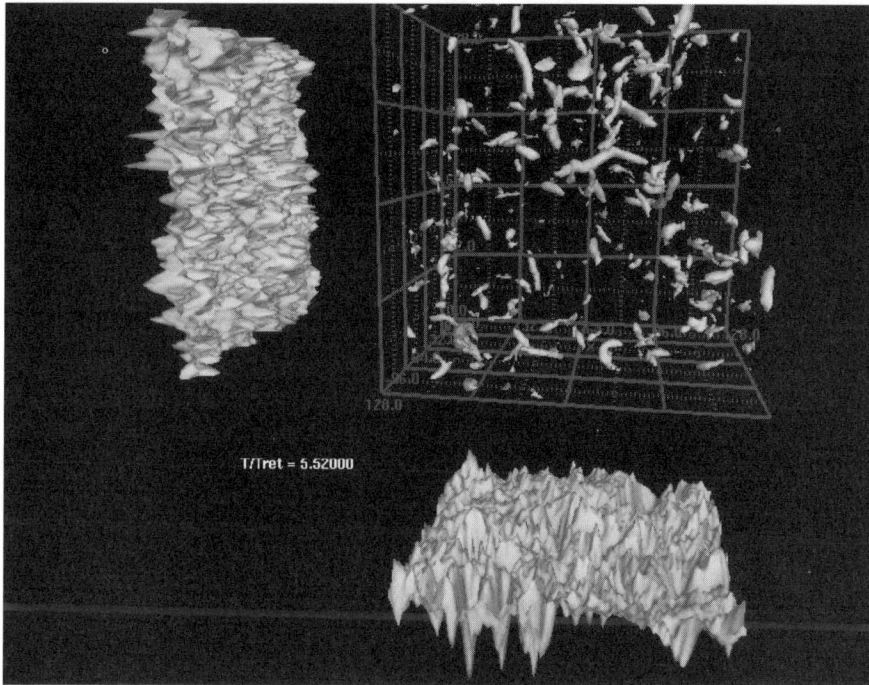

Figure 2.6. Isotropic decaying turbulence. Fixed view at $t = 6$ of pressure in Animation 2-1; $k_i = 4$; 128^3 modes.

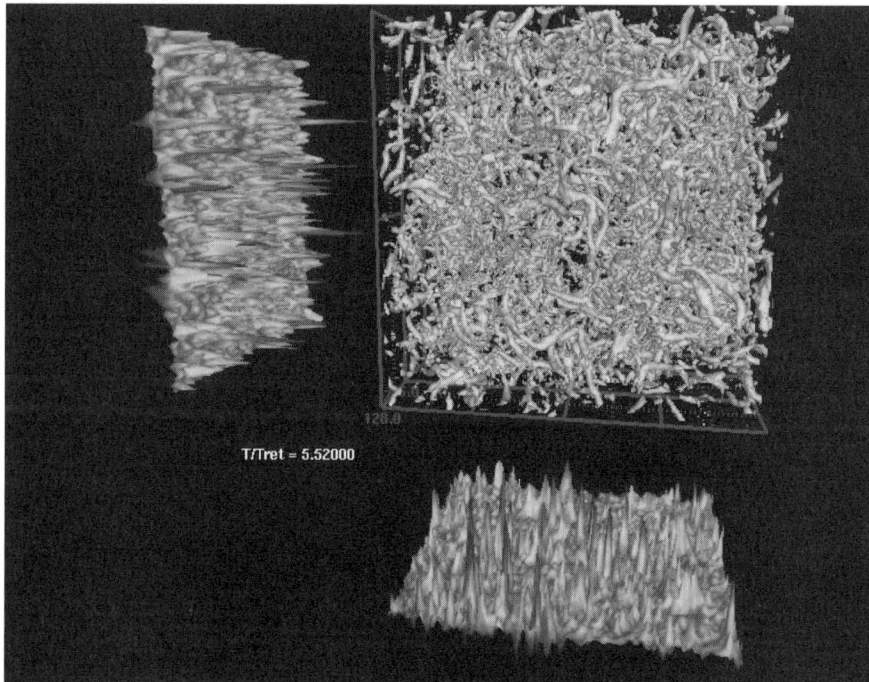

Figure 2.7. Isotropic decaying turbulence. Fixed view at $t = 6$ in the animation of Q isosurfaces at a given positive threshold.

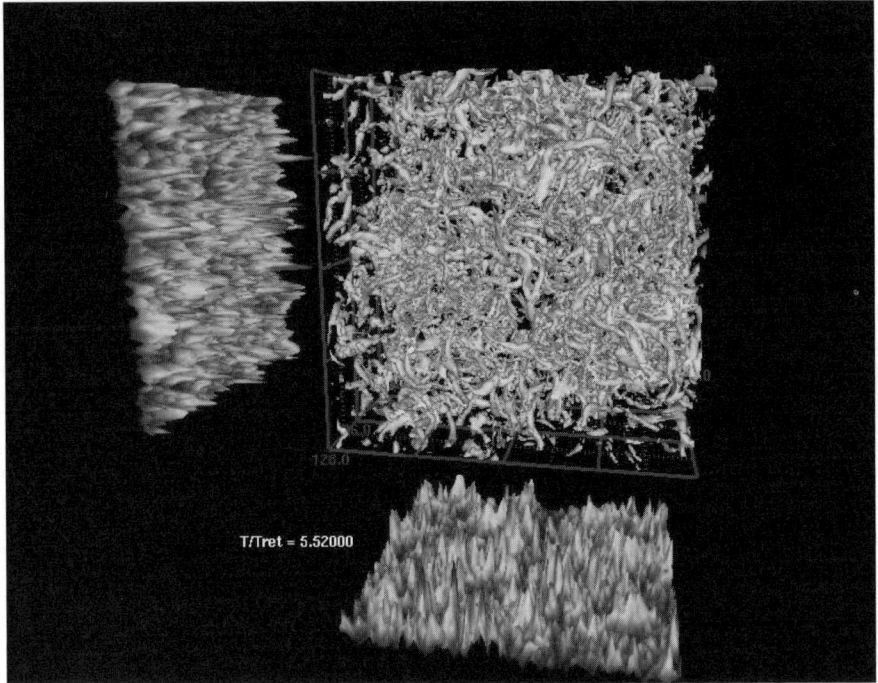

Figure 2.8. Isotropic decaying turbulence. Fixed view at $t = 6$ in the animation of the vorticity modulus.

When small-scale turbulence has developed, everybody seems to agree on the average tube length, which is the integral scale l. It is currently uncertain whether the diameter scales on the dissipative scale or on the Taylor microscale. Indeed, and if we interpret the vortices as resulting from the rollup of local vortex sheets, it is the Taylor microscale that should prevail, for it may be interpreted as proportional to some average local mixing-layer thickness within the flow. However, it is possible that vortex stretching (which seems to occur in the movie if there are no threshold-related artificial effects) diminishes the tubes' diameter up to the Kolmogorov scale. If fact, this strongly anisotropic vortex topology is very far from the quite naive spherical eddies considered in the popular folklore of Taylor–Richardson–Kolmogorov cascades.

A last remark on the structure of isotropic turbulence at small scales is in order: LES cannot of course give access to the smallest scales because they have been filtered out. However, the resolved motions display geometric features that resemble the fractals popularized by Mandelbrot [194]. Advanced multifractal studies of three-dimensional isotropic turbulence are presented in Frisch [103]. The multifractal character of turbulence might be at the origin of departures from Kolmogorov's 1941 laws [145] concerning the velocity structure functions of high order.

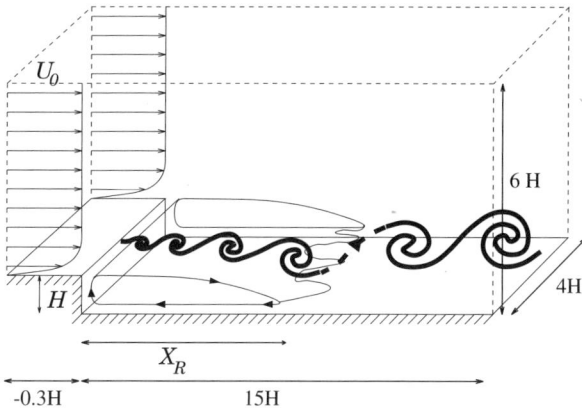

Figure 2.9. Schematic view of the backward-facing step. (From Delcayre [69].)

2.3.2 Backward-facing step

We now present LES results of a uniform-density flow above a straight backward-facing step. The code used is TRIO-VF, a tool developed for industrial applications of turbulence modeling and LES by the Commissariat à l'Energie Atomique (CEA) in France. (Details of this code will be given in Chapter 6.) The model used is the selected structure function (SSF) model (see Chapter 4). An animation of the simulation is presented on the enclosed CD-ROM (Animation 2-4). Figure 2.9 shows a schematic view of the flow. The step height is H, the expansion ratio is 1.2, and the Reynolds number is $U_0 H/\nu = 5,100$ as in the configuration studied experimentally (Jovic and Driver [138]) and numerically by Le et al. [159] using DNS. A free-slip boundary condition is used on the upper boundary. This is well justified based on laboratory experiments consisting of a double-expansion channel with potential laminar flow in its central part.

At the inlet, Spalart's [273] mean turbulent boundary layer velocity profile is imposed. A small three-dimensional white-noise perturbation regenerated at each time step is superposed to the latter. One assumes periodicity in the spanwise direction, and there is an outflow boundary condition of the Sommerfeld type, where the quantities are transported following a fictitious "tangential" wave-phase velocity (Orlanski [223]). We have determined that the latter is very good for letting the coherent vortices get out of the computational domain without any distortion. Animation 2-4 displays the following vortex dynamics: Quasi-two-dimensional Kelvin–Helmholtz-type vortices are shed behind the step, resulting from the instability of the upstream vortex sheet. Then they are subject to dislocations (helical pairings) and transform into a field of large, staggered archlike vortices, which impact the lower wall and are carried away downstream with their legs lying longitudinally close to the wall and

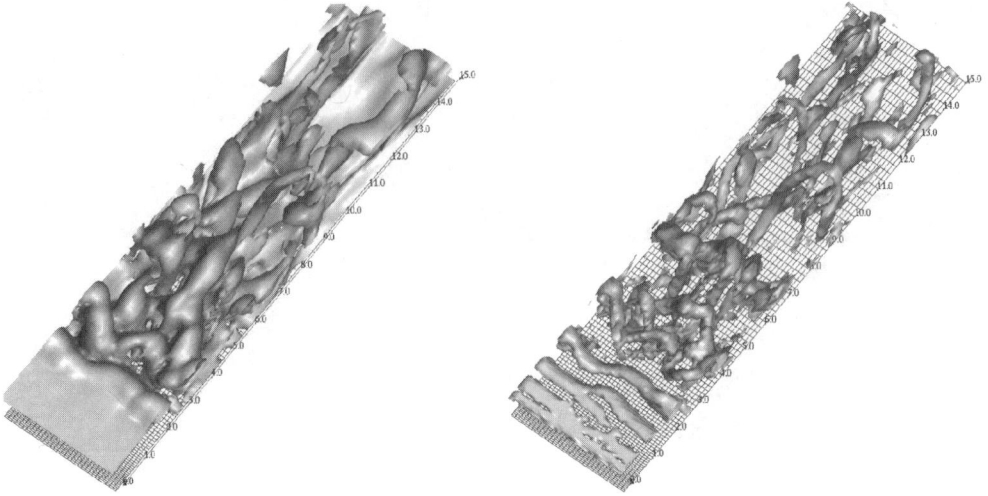

Figure 2.10. Incompressible backward-facing step; visualization of coherent vortices using high-vorticity modulus (left) and positive Q (right) isosurfaces. (From Delcayre [69].)

progressively raising away from the wall, as a result of the aforementioned self-induction of dipoles.

Figure 2.10 (left), presenting isovorticity maps, does show the breakdown of the vortex sheet into large staggered Λ vortices. Figure 2.10 (right) presents iso-Q maps and indicates the same vortex events as for the vorticity, but the vortices are thinner[6] and the upstream vortex sheet has been erased. In fact, the vortices in the movie are colored both by Q and by the longitudinal vorticity (positive, gray, negative, dark), so that their right and left legs are colored, respectively, in these two colors. Plots based on λ_2 (not shown here) are similar to Q maps. Finally, isobaric surfaces (Figure 2.11) are misleading in this case because they seem to indicate a large quasi-two-dimensional vortex at the level of reattachment, whereas it is simply an erroneous reconnection of the tips of the big Λs.

In a more recent LES study of the same step at Mach 0.3, Lesieur et al. [176] compare flows resulting from two sets of upstream conditions:

A. a mean velocity profile corresponding to Spalart's boundary-layer DNS [273] perturbed by a weak three-dimensional white noise, and
B. a more realistic, time-dependent velocity field (precursor calculation) generated through an extension to the compressible case of the method developed by Lund et al. [192].

In case B, the upstream boundary layer contains quasi-longitudinal vortices propagating before the step, as can be seen on the bottom of Figures 2.12 and 2.13.

---

[6] This might be due to an ill-chosen threshold.

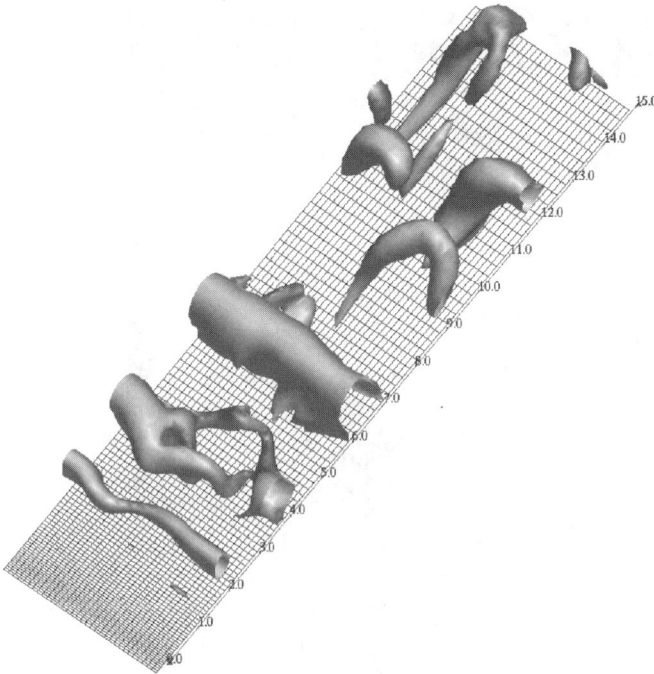

Figure 2.11. Incompressible backward-facing step showing low-pressure isosurfaces. (From Delcayre [69].)

The first grid point in the direction normal to the wall is at a distance of 1.3 in wall units relative to the upstream turbulent boundary layer. Periodicity is assumed in the spanwise direction, and the boundary conditions at the top and the exit of the domain are nonreflective. This makes a difference with respect to the DNS of [159] where free-slip conditions are taken. Figures 2.12 and 2.13 show Q isosurfaces (with a threshold of $0.6U_0^2/H^2$) for the two classes of upstream conditions. In Figure 2.12 (top), one sees the regular shedding of straight quasi-two-dimensional Kelvin–Helmholtz vortices, which appear at a distance of $1.5-2H$ downstream of the step. They undergo helical pairing and transform into big Λ vortices (arch vortices) that impinge on the lower wall and are carried away from the step. Figure 2.12 (bottom) shows qualitatively the same events, but vortices appear very close to the step, and the flow is much more three-dimensional. Helical pairing seems to be triggered by the passage of upstream longitudinal vortices passing above the step. The side views of Figure 2.13 confirm that the flow reattaches sooner in this case than in the noisy case. This is confirmed by the determination of the reattachment length, which is $5.80H$ for condition A and $5.29H$ for condition B. The latter value is different from the value of $\approx 6.1H$ found in [159] with equivalent upstream conditions. This discrepancy may be attributed to the differences in the boundary conditions above and downstream of the computational domain. Animations 2-5 and 2-6 illustrate the two types of simulations (noisy and precursor).

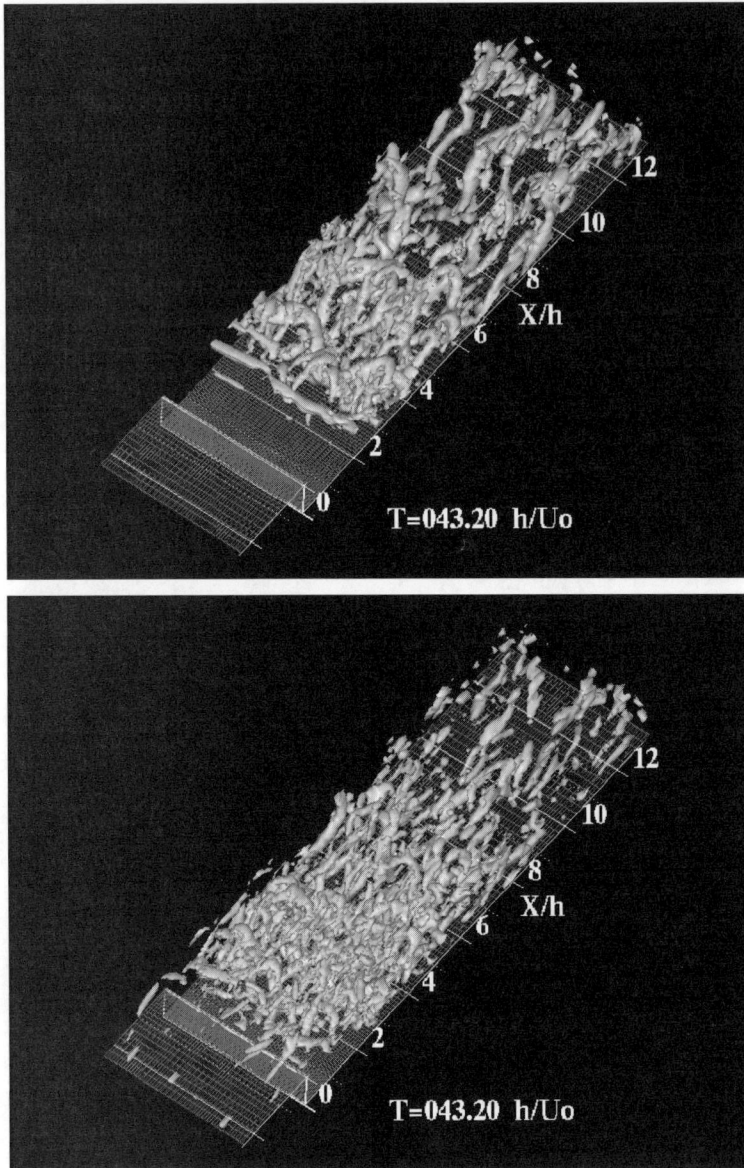

Figure 2.12. Perspective view of the Mach 0.3 backstep. (Top) Noised upstream velocity. (Bottom) Precursor upstream velocity. (From Danet [60].)

2.4 Coherent Structures

We define coherent structures in a much more general way than coherent vortices as structures displaying at a given time some spatial organization in space. In this respect, low- and high-speed streaks observed close to the wall in turbulent boundary layers, channels, and pipes are coherent structures.

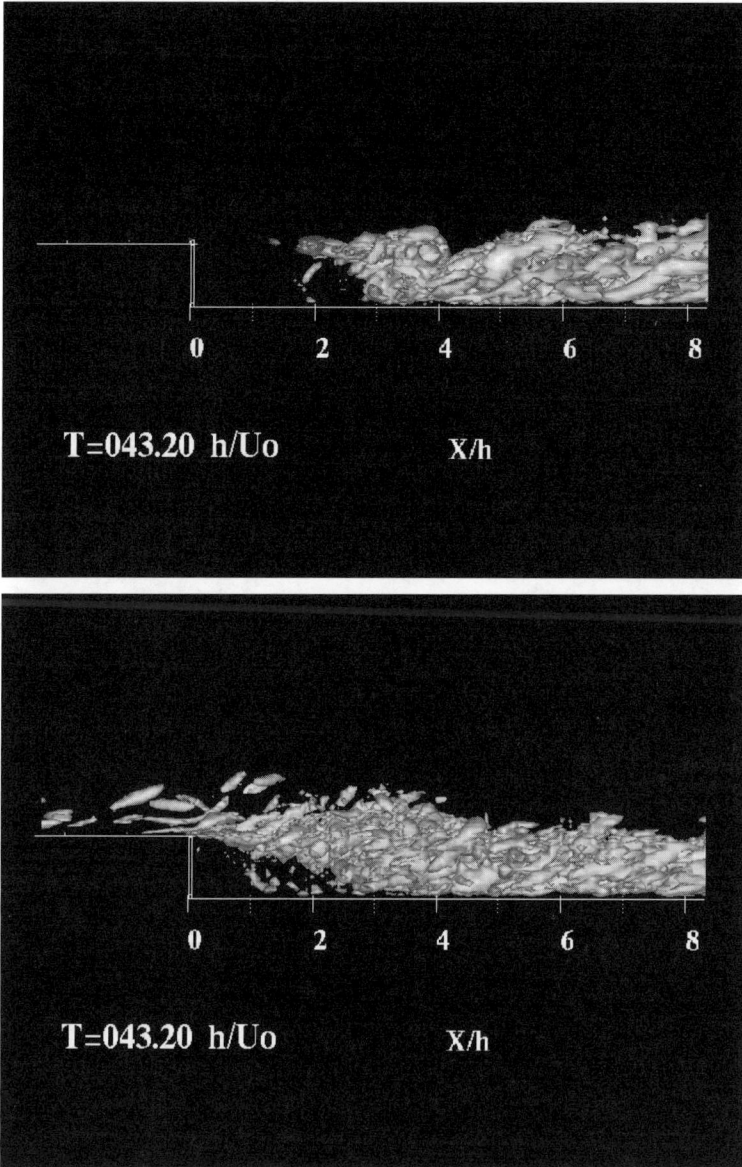

Figure 2.13. Side view of the Mach 0.3 backstep. (Top) Noised upstream velocity. (Bottom) Precursor upstream velocity. (From Danet [60].)

2.5 Animations

Animation 2-1: LES of decaying isotropic turbulence. Low macropressure isosurfaces from $t = 0$ to 15 initial large-eddy turnover times; $k_i = 4$. (Film 2-1.mpg; courtesy P. Begou.)

Animation 2-2: Same as Animation 2-1 for positive Q isosurfaces. (Film 2-2.mpg; courtesy P. Begou.)

Animation 2-3: Same as Animation 2-1 for vorticity norm isosurfaces. (Film 2-3.mpg; courtesy P. Begou.)

Animation 2-4: Shedding of arch vortices downstream of a straight backward-facing step. Vortices are visualized by Q isosurfaces colored by longitudinal vorticity. (Film 2-4.mpg; courtesy F. Delcayre.)

Animation 2-5: Noisy backward-facing step at Mach 0.3 showing Q isosurfaces. (Film 2-5.mpg; courtesy A. Danet.)

Animation 2-6: Backward-facing step at Mach 0.3 showing precursor inflow and Q isosurfaces. (Film 2-6.mpg; courtesy A. Danet.)

3 LES formalism in physical space

This chapter deals with an incompressible flow whose density is conserved with the fluid motion, which implies the continuity equation $\vec{\nabla}.\vec{u} = 0$. Then ρ may either be uniform or have a mean variation taken into account through Boussinesq's approximation (see Lesieur [170], Chapter II).

3.1 LES equations for a flow of constant density

To begin with, let us consider a numerical solution of the Navier–Stokes equations with constant density ρ_0 carried out in physical space, using finite-difference or finite-volume methods. Let Δx be a scale characteristic of the grid mesh. To eliminate the subgrid scales, we introduce a filter of width Δx. Mathematically, the filtering operation corresponds to the convolution of any quantity $f(\vec{x}, t)$ of the flow by the filter function $G_{\Delta x}(\vec{x})$ in the form

$$\bar{f}(\vec{x}, t) = \int f(\vec{y}, t) G_{\Delta x}(\vec{x} - \vec{y}) d\vec{y} = \int f(\vec{x} - \vec{y}, t) G_{\Delta x}(\vec{y}) d\vec{y}, \quad (3.1)$$

and the subgrid-scale field is the departure of the actual flow with respect to the filtered field:

$$f = \bar{f} + f'. \quad (3.2)$$

Since Δx is for the moment assumed constant,[1] it is easy to show that the space and time derivatives commute with the filtering operator.

We use a Cartesian system of coordinates. Let us first write the linear-momentum equations as

$$\frac{\partial u_i}{\partial t} + \frac{\partial}{\partial x_j}(u_i u_j) = -\frac{1}{\rho_0}\frac{\partial p}{\partial x_i} + \frac{\partial}{\partial x_j}(2\nu S_{ij}), \quad (3.3)$$

---

[1] This assumption may pose some problems later if irregular computational grids are used, although we never said that Δx was identical to the grid mesh.

where S_{ij} is the deformation tensor already defined. The filtered momentum equations are exactly

$$\frac{\partial \bar{u}_i}{\partial t} + \frac{\partial}{\partial x_j}(\bar{u}_i \bar{u}_j) = -\frac{1}{\rho_0}\frac{\partial \bar{p}}{\partial x_i} + \frac{\partial}{\partial x_j}(2\nu \bar{S}_{ij} + T_{ij}), \qquad (3.4)$$

where

$$T_{ij} = \bar{u}_i \bar{u}_j - \overline{u_i u_j} \qquad (3.5)$$

is the subgrid-stresses tensor responsible for momentum exchanges between the subgrid and the filtered scales. The filtered continuity equation is

$$\frac{\partial \bar{u}_j}{\partial x_j} = 0. \qquad (3.6)$$

Let us consider now the mixing of a scalar (such as temperature or density) of molecular diffusivity κ transported by the flow and satisfying the equation

$$\frac{\partial \rho}{dt} + \frac{\partial}{\partial x_j}(\rho u_j) = \frac{\partial}{\partial x_j}\left\{\kappa \frac{\partial \rho}{\partial x_j}\right\}. \qquad (3.7)$$

The filtered scalar equation is then

$$\frac{\partial \bar{\rho}}{dt} + \frac{\partial}{\partial x_j}(\bar{\rho}\bar{u}_j) = \frac{\partial}{\partial x_j}\left\{\kappa \frac{\partial \bar{\rho}}{\partial x_j} + T_j^{(\rho)}\right\}, \qquad (3.8)$$

where

$$T_j^{(\rho)} = \bar{\rho}\bar{u}_j - \overline{\rho u_j} \qquad (3.9)$$

is the subgrid scalar flux. T_{ij} and $T_j^{(\rho)}$ can be written as

$$T_{ij} = -\left(\overline{u_i' u_j'} + \overline{\bar{u}_i u_j'} + \overline{u_i' \bar{u}_j} + \overline{\bar{u}_i \bar{u}_j} - \bar{u}_i \bar{u}_j\right), \qquad (3.10)$$

$$T_j^{(\rho)} = -\left(\overline{\rho' u_j'} + \overline{\bar{\rho} u_j'} + \overline{\rho' \bar{u}_j} + \overline{\bar{\rho}\bar{u}_j} - \bar{\rho}\bar{u}_j\right). \qquad (3.11)$$

In Eq. (3.10), $-\overline{u_i' u_j'}$ is a Reynolds-stress-like term, $-(\overline{\bar{u}_i u_j'} + \overline{u_i' \bar{u}_j})$ is called the Clark term (Clark et al. [46]), and $\bar{u}_i \bar{u}_j - \overline{\bar{u}_i \bar{u}_j}$ is the Leonard tensor [163]. The latter is explicit in the sense that it is defined in terms of the filtered field, and it has been used in scale-similarity models to provide information on the subgrid stresses (see Chapter 6). Leonard's stresses are also a major ingredient of the so-called Germano's identity for the dynamic approach in physical space.

These subgrid-scale tensors and fluxes need of course to be modeled.

3.2 LES Boussinesq equations in a rotating frame

We give now the LES equations corresponding to the Navier–Stokes equations within the Boussinesq approximation in a Cartesian frame of reference rotating with a constant angular velocity Ω about the x_3 axis. The momentum equation is

$$\frac{\partial \bar{u}_i}{\partial t} + \frac{\partial}{\partial x_j}(\bar{u}_i \bar{u}_j) = -\frac{1}{\rho_0}\frac{\partial \bar{p}}{\partial x_i} + \frac{\partial}{\partial x_j}(2\nu \bar{S}_{ij} + T_{ij}) + 2\epsilon_{ij3}\Omega \bar{u}_j + g_i\delta_{i3}\frac{\bar{\rho}}{\rho_0},$$
(3.12)

where g_i are the gravity components, and ρ_0 is the average of the density on the thickness of the fluid layer.

   This equation comes from the filtering of a particular version of the Boussinesq equations, which is valid for both a liquid and a perfect gas (see [170], p. 45), where p is the static pressure. For a liquid, $\bar{\rho}$ satisfies

$$\frac{\partial \bar{\rho}}{dt} + \frac{\partial}{\partial x_j}(\bar{\rho}\bar{u}_j) = \frac{\partial}{\partial x_j}\left\{\kappa \frac{\partial \bar{\rho}}{\partial x_j} + T_j^{(\rho)}\right\}.$$
(3.13)

For a perfect gas, one can show within the Boussinesq approximation that

$$\frac{\bar{\rho}}{\rho_0} = \frac{\bar{\rho}}{\bar{\Theta}}\frac{\bar{\Theta}}{\rho_0} = -\beta\bar{\Theta},$$
(3.14)

with

$$\beta = -\frac{1}{\rho_0}\frac{\partial \bar{\rho}}{\partial \bar{\Theta}}.$$
(3.15)

This requires that

$$\frac{\partial \bar{\rho}}{\partial \bar{\Theta}} = \frac{\bar{\rho}}{\bar{\Theta}}.$$
(3.16)

Such a result holds because relative increments of temperature, potential temperature, and minus density are, within the Boussinesq approximation, equal and much larger than the relative pressure increment (see [170], p. 46). If we assume that the volumetric expansion coefficient, β, is constant, then $\bar{\Theta}$ satisfies Eq. (3.13), and the momentum equation becomes

$$\frac{\partial \bar{u}_i}{\partial t} + \frac{\partial}{\partial x_j}(\bar{u}_i \bar{u}_j) = -\frac{1}{\rho_0}\frac{\partial \bar{p}}{\partial x_i} + \frac{\partial}{\partial x_j}(2\nu \bar{S}_{ij} + T_{ij}) + 2\epsilon_{ij3}\Omega \bar{u}_j - \beta g_i\delta_{i3}\bar{\Theta},$$
(3.17)

where T_{ij} is defined by Eq. (3.5) and $T_j^{(\rho)}$ by Eq. (3.9).

   Although, ρ and Θ are still scalars transported by the flow [as in Eq. (3.7)], they are not passive because they react with the velocity field, through gravity, in the momentum equation. In fact, this system of equations is very useful for studying stably stratified or thermally convective rotating flows. The equations have been used in the latter case by Schmidt and Schumann [258] for a study of coherent structures in a thermally convective boundary layer. As already stressed, Schmidt and Schumann used a one-point closure modeling point of view within second-order closure transport equations with the coefficients being determined from a spectral inertial-range analysis. This study displays evidence of small-scale plumes as well as the existence of large-scale cold updraufts and warm downdraufts at the top of the layer. Analogous LES work was also carried out by Mason [199].

We will use Boussinesq LES equations in Chapter 8 for atmospheric storms and oceanic deep-water formation studies.

3.3 Eddy viscosity and diffusivity assumption

By analogy with what is done in the framework of Reynolds equations for the ensemble-averaged equations, the subgrid-scale tensors are in most of the cases expressed in terms of eddy viscosity and diffusivity coefficients in the form

$$T_{ij} = 2\nu_t(\vec{x}, t)\,\bar{S}_{ij} + \frac{1}{3} T_{ll}\,\delta_{ij}, \qquad T_j^{(\rho)} = \kappa_t(\vec{x}, t)\,\frac{\partial \bar{\rho}}{\partial x_j}. \qquad (3.18)$$

Then the LES equations for a flow of uniform density without rotation can be written as

$$\frac{\partial \bar{u}_i}{\partial t} + \frac{\partial}{\partial x_j}(\bar{u}_i \bar{u}_j) = -\frac{1}{\rho_0}\frac{\partial \bar{P}}{\partial x_i} + \frac{\partial}{\partial x_j}\left\{(\nu + \nu_t)\left(\frac{\partial \bar{u}_i}{\partial x_j} + \frac{\partial \bar{u}_j}{\partial x_i}\right)\right\}, \qquad (3.19)$$

$$\frac{\partial \bar{\rho}}{\partial t} + \frac{\partial}{\partial x_j}(\bar{\rho}\bar{u}_j) = \frac{\partial}{\partial x_j}\left\{(\kappa + \kappa_t)\frac{\partial \bar{\rho}}{\partial x_j}\right\}, \qquad (3.20)$$

where

$$\bar{P} = \bar{p} - \frac{1}{3}\rho_0 T_{ll} \qquad (3.21)$$

is a modified pressure (macropressure), which can be determined with the aid of the filtered continuity equation.

Several questions are in fact posed. The first one is how to determine the eddy viscosity ν_t and the corresponding turbulent Prandtl number

$$Pr_t = \frac{\nu_t}{\kappa_t}, \qquad (3.22)$$

and the second one concerns the validity of the eddy-viscosity assumption itself. Indeed, it is based on an analogy with Newtonian fluids, which is certainly not fulfilled here. Let us briefly discuss this point. Molecular viscosity ν characterizes the momentum exchanges for a "macroscopic" fluid parcel with the surrounding fluid owing to molecular diffusion across its interface. Here, one assumes a wide separation between macroscopic and microscopic scales,[2] and it is this separation that allows us to calculate these molecular exchange coefficients using kinetic theories of liquids or gases in which molecules are assumed to follow some sort of Gaussian random walk. No such scale separation exists in the LES problem, where one observes in general a distribution of energy (kinetic-energy spectrum) continuously decreasing from the energetic to the smallest dissipative scales even in inflectional shear flows with vigorous coherent vortices. Because the cutoff scale Δx lies in the middle of this spectrum, there is obviously no spectral gap at this level. Furthermore, trajectories of fluid parcels are very far from a random walk, for they may be either trapped around a vortex or strained in stagnation regions between vortices.

We believe therefore that the lack of a spectral gap is the major drawback of the eddy-viscosity assumption in physical space and is responsible for the fact that numerous numerical and even experimental a priori tests (see, e.g., Clark et al. [46] and Liu et al. [188]) invalidate relations (3.18): When a low-pass filter is, for instance, applied to DNS results, one can calculate explicitly the subgrid-stress tensors and correlate them to the filtered deformation. The correlation found is very poor and is of the order of 0.1 instead of 1. This justifies the development of models going beyond the classical eddy-viscosity concept – for example, the spectral eddy viscosity (see Chapter 4) and also the models presented in Chapter 6. However, LES results based on classical eddy viscosities in physical space derived from Smagorinsky or structure-function models may give very good results, as will be seen later, from the point of view of vortex dynamics and statistical predictions.

Another problem concerns the macropressure that has been introduced, for it contains the unknown trace of the subgrid-stresses tensor. This is fine if one is not interested in the exact value of the static pressure. If, however, the latter is needed,[3] then it is necessary to model the trace T_{ll}. A similar problem arises in LES of compressible turbulence, for which the same type of macropressure will be introduced (see Chapter 7).

Notice finally that the use of subgrid models in a rotating frame poses realizability problems; these are discussed by Horiuti [127] and Domaradzki and Horiuti [71].

---

[2] As already emphasized in Chapter 1, this is valid except for hypersonic flows at very high Mach numbers at which the two scales become of the same order and the equations of motion have to be replaced by Boltzmann equations.

[3] For instance, in cavitation studies, where cavitation occurs when the pressure goes below a given threshold and may cause severe damage by bubble implosion to the material in contact with the fluid.

3.3.1 Smagorinsky's model

As already pointed out, the most widely used eddy-viscosity model was proposed by Smagorinsky [269]. He introduced an eddy viscosity that was supposed to take into account subgrid-scale dissipation through a Kolmogorov $k^{-5/3}$ cascade. Smagorinsky's model is an adaptation of Prandtl's mixing-length theory to subgrid-scale modeling. Prandtl assumes that the eddy viscosity arising in RANS equations is proportional to a turbulence characteristic scale (the mixing length) multiplied by a turbulence characteristic velocity. In the same way, Smagorinsky supposes that the LES eddy viscosity is proportional to the subgrid-scale characteristic length Δx and to a characteristic subgrid-scale velocity

$$v_{\Delta x} = \Delta x \, |\bar{S}|, \tag{3.23}$$

based on the second invariant of the filtered-field deformation tensor

$$|\bar{S}| = \sqrt{2\bar{S}_{ij}\bar{S}_{ij}}. \tag{3.24}$$

Thus, Smagorinsky's eddy viscosity is

$$v_t = (C_S \Delta x)^2 |\bar{S}|. \tag{3.25}$$

The constant may be calculated in isotropic turbulence, as was done by Lilly [184]. Let us assume that $k_C = \pi/\Delta x$, the cutoff wavenumber in Fourier space, lies within a $k^{-5/3}$ Kolmogorov cascade

$$E(k) = C_K \langle \epsilon_{sm} \rangle^{2/3} k^{-5/3}. \tag{3.26}$$

One can show for a sharp filter in Fourier space that the dissipation rate of the resolved kinetic energy is

$$2v_t \langle \bar{S}_{ij}\bar{S}_{ij} \rangle = \int_0^{k_C} 2v_t k^2 E(k) dk, \tag{3.27}$$

which yields

$$\langle \bar{S}_{ij}\bar{S}_{ij} \rangle = \frac{3}{4} C_K \langle \epsilon_{sm} \rangle^{2/3} \left(\frac{\pi}{\Delta x}\right)^{4/3}$$

and

$$\langle \epsilon_{sm} \rangle = \left(\frac{4}{3}\right)^{3/2} C_K^{-3/2} \left(\frac{\pi}{\Delta x}\right)^{-2} \langle \bar{S}_{ij}\bar{S}_{ij} \rangle^{3/2}. \tag{3.28}$$

Another expression of $\langle \epsilon_{sm} \rangle$ may be obtained through an assumption of local equilibrium, leading to

$$\langle \epsilon_{sm} \rangle = \langle 2v_t \bar{S}_{ij}\bar{S}_{ij} \rangle, \tag{3.29}$$

which, using Smagorinsky's expression for ν_t, can be expressed as

$$\langle \epsilon_{sm} \rangle = 2^{3/2} C_S^2 \Delta x^2 \left\langle \left(\overline{D}_{ij} \overline{D}_{ij} \right)^{3/2} \right\rangle. \tag{3.30}$$

Equating the expressions of Eqs. (3.28) and (3.30) for $\langle \epsilon_{sm} \rangle$, we obtain

$$C_S = \frac{1}{\pi} \left(\frac{3C_K}{2} \right)^{-3/4} \sqrt{\frac{\left(\overline{D}_{ij} \overline{D}_{ij} \right)^{3/2}}{\left\langle \left(\overline{D}_{ij} \overline{D}_{ij} \right)^{3/2} \right\rangle}}. \tag{3.31}$$

For a Gaussian field, the coefficient under the square root on the r.h.s. of Eq. (3.31) is equal to one, and Smagorinsky's constant will be approximated by

$$C_S \approx \frac{1}{\pi} \left(\frac{3C_K}{2} \right)^{-3/4}. \tag{3.32}$$

This yields $C_S \approx 0.18$ for a Kolmogorov constant of 1.4. This value proves to give acceptable results for LES of isotropic turbulence. However, most researchers prefer $C_S = 0.1$, which represents a reduction by nearly a factor of 4 in the eddy viscosity. At this value, Smagorinsky's model behaves reasonably well for free-shear flows and for wall flows with wall laws, as in the channel LES of Moin and Kim [210]. Let us mention also the work of Breuer and Jovivi [30] on incompressible LES of a separated flow around a two-dimensional airfoil at an incidence angle of 18°. Such a calculation also displays quite nicely the three-dimensional vortical structure of such a flow, including a Karman street. A valuable assessment study of various subgrid models applied to some well-defined test cases for which experiments exist has been carried out by Rodi et al. [244].

If one does not want to play with C_S at the boundary, Smagorinsky's model is too dissipative in the presence of a wall; moreover, it does not work in particular for transition in a boundary-layer developing on a flat plate: It artificially relaminarizes the flow if the upstream perturbation is not high enough. This is due to the heavy influence in the eddy viscosity of the velocity gradient in the direction normal to the wall and to an improper behavior of the model at the wall, which we are going to discuss in more detail. A good review of this problem is given by Meneveau and Katz [202]. Let us assume that the velocity components close to the wall may be expanded in Taylor series as

$$u(x, y, z, t) = a_1(x, z, t)y + a_2(x, z, t)y^2 + a_3(x, z, t)y^3 + \cdots,$$

$$v(x, y, z, t) = b_1(x, z, t)y + b_2(x, z, t)y^2 + b_3(x, z, t)y^3 + \cdots,$$

$$w(x, y, z, t) = c_1(x, z, t)y + c_2(x, z, t)y^2 + c_3(x, z, t)y^3 + \cdots.$$

Using the continuity equation, we obtain up to the first order

$$\frac{\partial a_1}{\partial x}y + b_1 + 2b_2 y + \frac{\partial c_1}{\partial z}y = 0,$$

and hence $b_1 = 0$. Introducing wall units (recalled in Chapter 5), we then have for $y^+ \to 0$

$$u \propto y^+; \quad w \propto y^+; \quad v \propto \left(y^+\right)^2. \tag{3.33}$$

Let us consider the component $T_{12} = \overline{u}\,\overline{v} - \overline{uv}$ of the subgrid-scale tensor. Close to the wall, we have

$$T_{12} \propto \left(y^+\right)^3. \tag{3.34}$$

The prediction of Smagorinsky's model for this component is

$$T_{12} = 2\nu_t(\vec{x}, t)\overline{S}_{12} = 2\left(C_S \Delta x\right)^2 |\overline{S}|\overline{S}_{12} \approx 2\left(C_S \Delta x\right)^2 \left(\frac{\partial u}{\partial y}\right)^2. \tag{3.35}$$

Close to the wall, $\partial u/\partial y$ is finite, and the Smagorinsky model yields therefore a finite value for T_{12}. This justifies Smagorinsky's dynamic approach, which will be presented now, where the constant is dynamically adjusted to the flow conditions.

As noted by Lilly [183, 185, 187], the eddy viscosity given by Eq. (3.23) may be expressed by taking a different value for the velocity $v_{\Delta x}$ through Kolmogorov's relation corresponding to Eq. (1.42). This yields in the latter case

$$\nu_t \sim \epsilon^{1/3} \Delta x^{4/3} \sim \epsilon^{1/3} k_C^{-4/3}, \tag{3.36}$$

an expression that was also used by Schmidt and Schumann [258] under a slightly different form based on the rms kinetic energy. A review of these models, which have analogies with the structure-function model, may be found in Muchinski [215].

3.3.2 Dynamic Smagorinsky model

The underlying principle of the dynamic model is to extract information concerning a given eddy-viscosity model via a double filtering in physical space (Germano [109]). Most of the historical developments have been done with Smagorinsky's model, but the dynamic procedure applies in fact to other types of eddy viscosities such as those used in the structure-function model. The following presentation is very close to that in Lesieur [170].

We start with a regular LES corresponding to a "bar-filter" of width Δx, an operator associating a function (which may be a scalar, a vector, or a tensor) $\bar{f}(\vec{x}, t)$ with the function $f(\vec{x}, t)$. We then define a second "test filter" *tilde* of larger width $\alpha \Delta x$ (for instance $\alpha = 2$), associating $\tilde{f}(\vec{x}, t)$ with $f(\vec{x}, t)$. We then have two filter operators, *bar* and *tilde*, that apply to functions, the product being *tilde* o *bar*.[4] This product, applied to $f(\vec{x}, t)$, means that we first apply to f the *bar* filter (to yield $\bar{f}$) and then the *tilde* filter to obtain $\widetilde{\bar{f}}$. Let us first apply this filter product to the Navier–Stokes equation (with constant density). The subgrid-scale tensor of the field $\widetilde{\bar{u}}_i$ is obtained from Eq. (3.5) with the replacement of the filter *bar* by the double filter:

$$\mathcal{T}_{ij} = \widetilde{\bar{u}}_i \widetilde{\bar{u}}_j - \widetilde{\overline{u_i u_j}}. \tag{3.37}$$

We consider now the field $\bar{u}_i$ per se[5] and evaluate the resolved turbulent stresses obtained by application of the *tilde* filter. We can then write

$$\mathcal{L}_{ij} = \widetilde{\bar{u}}_i \widetilde{\bar{u}}_j - \widetilde{\bar{u}_i \bar{u}_j}. \tag{3.38}$$

We now apply the *tilde* filter to Eq. (3.5), which leads to

$$\widetilde{T}_{ij} = \widetilde{\bar{u}_i \bar{u}_j} - \widetilde{\overline{u_i u_j}}. \tag{3.39}$$

Adding Eqs. (3.38) and (3.39) and using Eq. (3.37), we obtain

$$\mathcal{L}_{ij} = \mathcal{T}_{ij} - \widetilde{T}_{ij}. \tag{3.40}$$

This expression is called Germano's identity. On the r.h.s., $\mathcal{T}_{ij}$ and $\widetilde{T}_{ij}$ have to be modeled, whereas the l.h.s. $\mathcal{L}_{ij}$ (the resolved stresses) can be explicitly calculated by applying the *tilde* filter to $\bar{u}_i$.

We use Smagorinsky's model expression for the subgrid stresses related to the *bar* filter and "tilde-filter" it to get

$$\widetilde{T}_{ij} - \frac{1}{3}\widetilde{T}_{ll}\,\delta_{ij} = 2\widetilde{\mathcal{A}_{ij}}C, \tag{3.41}$$

where the constant C is equal to $2C_S^2$, and

$$\mathcal{A}_{ij} = (\Delta x)^2\, |\bar{S}|\bar{S}_{ij}.$$

We now have to determine $\mathcal{T}_{ij}$, the stress resulting from the filter product. This is again obtained using the Smagorinsky model, which yields

$$\mathcal{T}_{ij} - \frac{1}{3}\mathcal{T}_{ll}\,\delta_{ij} = 2\mathcal{B}_{ij}C, \tag{3.42}$$

with

$$\mathcal{B}_{ij} = \alpha^2(\Delta x)^2\, |\widetilde{\bar{S}}|\,\widetilde{\bar{S}}_{ij}.$$

---

[4] The product is in the sense of product of operators.
[5] The field is considered as if it were the instantaneous field.

Subtracting Eq. (3.41) from Eq. (3.42) yields with the aid of Germano's identity

$$\mathcal{L}_{ij} - \frac{1}{3}\mathcal{L}_{ll}\,\delta_{ij} = 2\mathcal{B}_{ij}C - 2\widetilde{\mathcal{A}_{ij}C}.$$

This is a nice result relating the (unknown) model coefficient to the resolved stresses. However, there are some difficulties. First, one removes C from the filtering as if it were constant,[6] leading to

$$\mathcal{L}_{ij} - \frac{1}{3}\mathcal{L}_{ll}\,\delta_{ij} = 2C M_{ij},\tag{3.43}$$

with

$$M_{ij} = \mathcal{B}_{ij} - \widetilde{\mathcal{A}_{ij}}.$$

All the terms of Eq. (3.43) may now be determined with the aid of $\bar{\bar{u}}$. Unfortunately, there are five independent equations for only one variable C, and thus the problem is overdetermined. A first solution proposed by Germano [109] is to multiply Eq. (3.43) tensorially by $\bar{S}_{ij}$ to get

$$C = \frac{1}{2}\frac{\mathcal{L}_{ij}\bar{S}_{ij}}{M_{ij}\bar{S}_{ij}}\tag{3.44}$$

(and owing to incompressibility, $\bar{S}_{ii} = 0$). This provides finally a dynamical evaluation of $C(\vec{x}, t)$, which can be used in the LES of the *bar* field $\bar{u}$. However, problems still arise: In tests using channel-flow data obtained from DNS, Germano [109] showed that the denominator in Eq. (3.44) could locally vanish or become sufficiently small to yield computational instabilities. Lilly [186] chose to determine the value of C in Eq. (3.43) by a variational approach using a least-squares method, which gives

$$C = \frac{1}{2}\frac{\mathcal{L}_{ij}M_{ij}}{M_{ij}^2}\tag{3.45}$$

and removes the indeterminacy of Eq. (3.43). However, and as discussed in Lesieur ([170], p. 405), the analysis of DNS data reveals that the C field predicted by the models (3.44) or (3.45) varies strongly in space and contains a significant fraction of negative values with a variance that may be ten times higher than the square mean. So, the removal of C from the filtering operation is not really justified and the model exhibits some mathematical inconsistencies. The possibility of negative C is an advantage of the model because it allows a sort of backscatter in physical space, but very large negative values of the eddy viscosity destabilize the numerical simulation, yielding

[6] This is in some way contradictory to the original aim of having a dynamic evaluation of C depending on space and time.

a nonphysical growth of the resolved-scales energy. The cure often adopted to avoid excessively large values of C consists in averaging the numerators and denominators of (3.44) and (3.45) over space, time, or both, thereby losing some of the conceptual advantages of the "dynamic" local formulation. Averaging over direction of flow homogeneity has been a popular choice, and good results have been obtained by Germano [109] and Piomelli and Balaras [239], who took averages in planes parallel to the walls in their channel-flow simulation. We remark that the same thing will be done with success when averaging the dynamic spectral eddy viscosity in channel-flow LES (see Chapter 5). Meneveau et al. [201] adopted a Lagrangian viewpoint and obtained good results in a dynamic Smagorinsky approach in which the constant C was averaged following the flow motion (see also Piomelli et al. [238] and Piomelli and Balaras [239]). This is in fact more physical as far as coherent vortices are concerned. It can be shown that the dynamic model gives a zero subgrid-scale stress at the wall, where L_{ij} vanishes, which is a great advantage with respect to the original Smagorinsky model; it also gives the proper asymptotic behavior near the wall.

4 Spectral LES for isotropic turbulence

We have seen that a major drawback of the eddy viscosity assumption in physical space is the nonexistence of a spectral gap between resolved and subgrid scales. This is an argument in favor of working in Fourier space, where we will see that the lack of a spectral gap may be dealt with in some sense.

4.1 Spectral eddy viscosity and diffusivity

We assume that the Navier–Stokes equation is written in Fourier space. This requires statistical homogeneity in the three directions of space, but we will see in the following how to handle flows with only one direction of inhomogeneity. Let $\hat{u}_i(\vec{k}, t)$ and $\hat{\rho}(\vec{k}, t)$ be the spatial Fourier transforms of, respectively, the velocity and passive-scalar fields introduced in Chapter 1. As already stressed, they are defined in the framework of generalized functions.[1] The filter consists of a sharp cutoff filter simply clipping all the modes larger than k_C, where $k_C = \pi/\Delta x$ is the cutoff wavenumber obtained when one uses a pseudo-spectral method in a given direction of periodicity.

We write the Navier–Stokes equation in Fourier space as

$$\frac{\partial}{\partial t}\hat{u}_i(\vec{k}, t) + [\nu + \nu_t(\vec{k}|k_C)]k^2\hat{u}_i(\vec{k}, t)$$

$$= -ik_m P_{ij}(\vec{k}) \int_{|\vec{p}|,|\vec{q}|<k_C}^{\vec{p}+\vec{q}=\vec{k}} \hat{u}_j(\vec{p}, t)\hat{u}_m(\vec{q}, t)d\vec{p}. \qquad (4.1)$$

The spectral eddy viscosity $\nu_t(\vec{k}|k_C)$ is defined by

$$\nu_t(\vec{k}|k_C)k^2\hat{u}_i(\vec{k}, t) = ik_m P_{ij}(\vec{k}) \int_{|\vec{p}|\text{or}|\vec{q}|>k_C}^{\vec{p}+\vec{q}=\vec{k}} \hat{u}_j(\vec{p}, t)\hat{u}_m(\vec{q}, t)d\vec{p}. \qquad (4.2)$$

---

[1] Discretized equivalents correspond to the discrete Fourier transforms of flows in spatially periodic domains.

At this point, it may not be positive or even real. The condition $\vec{p} + \vec{q} = \vec{k}$ is a "resonant-triad condition" resulting from the convolution coming from the Fourier transform of a product. The r.h.s. of Eq. (4.1) corresponds to a resolved transfer. A spectral eddy diffusivity for the passive scalar may be defined in the same way by writing the passive-scalar equation in Fourier space

$$\frac{\partial}{\partial t} \hat{\rho}(\vec{k}, t) + [\kappa + \kappa_t(\vec{k}|k_C)]k^2 \hat{\rho}(\vec{k}, t) = -i k_j \int_{|\vec{p}|,|\vec{q}|<k_C}^{\vec{p}+\vec{q}=\vec{k}} \hat{u}_j(\vec{p}, t)\hat{\rho}(\vec{q}, t) d\vec{p} \tag{4.3}$$

with

$$\kappa_t(\vec{k}|k_C)k^2 \hat{\rho}(\vec{k}, t) = i k_j \int_{|\vec{p}|_{\text{or}}|\vec{q}|>k_C}^{\vec{p}+\vec{q}=\vec{k}} \hat{u}_j(\vec{p}, t)\hat{\rho}(\vec{q}, t) d\vec{p}. \tag{4.4}$$

Expressions (4.2) and (4.4) give exact expressions of the eddy coefficients. They are, however, useless because they involve subgrid quantities. In fact, the eddy coefficients can be evaluated at the level of kinetic-energy and passive-scalar spectra evolution equations obtained with the aid of two-point closures of three-dimensional isotropic turbulence.

It is in this context that the concept of k-dependent eddy viscosity was first introduced by Kraichnan [147]. The spectral eddy diffusivity for a passive scalar was introduced by Chollet and Lesieur [42]. Kraichnan used the so-called test-field model. We work using a slightly different closure called the eddy-damped quasi-normal Markovian theory introduced by Orszag [224, 225] (see also André and Lesieur [6] and Lesieur [170] for details). We first briefly recall the main lines of this model.

4.2 EDQNM theory

In the EDQNM theory, which is easily manageable only in the case of isotropic turbulence, the fourth-order cumulants in the hierarchy of moments equations are supposed to relax the third-order moments linearly in the same qualitative way that the molecular viscosity does. Thus, a time θ_{kpq} characterizing this relaxation is introduced. The EDQNM gives for isotropic turbulence the following evolution equation for the kinetic-energy spectrum $E(k, t)$:

$$\left(\frac{\partial}{\partial t} + 2\nu k^2 \right) E(k, t)$$

$$= \iint_{\Delta_k} dp \, dq \, \theta_{kpq}(t) \frac{k}{pq} b(k, p, q) E(q, t)[k^2 E(p, t) - p^2 E(k, t)], \tag{4.5}$$

where the integration is carried out in the domain Δ_k of the (p, q) plane such that (k, p, q) can be the sides of a triangle and thus satisfy triangular inequalities. The nondimensional coefficient

$$b(k, p, q) = \frac{p}{k}(xy + z^3) \qquad (4.6)$$

is defined in terms of the cosines (x, y, z) of the interior angles of the triangle formed by the resonant triad $(\vec{k}, \vec{p}, \vec{q})$. The time $\theta_{kpq}(t)$ is given by

$$\theta_{kpq} = \frac{1 - e^{-[\mu_{kpq} + \nu(k^2 + p^2 + q^2)]t}}{\mu_{kpq} + \nu(k^2 + p^2 + q^2)} \qquad (4.7)$$

with

$$\mu_{kpq} = \mu_k + \mu_p + \mu_q$$

and

$$\mu_k = a_1 \left[\int_0^k p^2 E(p, t) dp \right]^{1/2}. \qquad (4.8)$$

The constant a_1 is adjusted in such a way that the kinetic-energy flux is equal to ϵ in a Kolmogorov cascade of infinite length, as done in André and Lesieur [6]. One finds $a_1 = 0.218\ C_K^{3/2}$. An analogous equation may be written for the passive-scalar spectrum $E_\rho(k, t)$ with a scalar transfer involving products $E E_\rho$. Let us present now some recent EDQNM results of decaying isotropic turbulence at high or very high Reynolds number obtained by Lesieur and Ossia [174]. The code used is the one developed by Lesieur and Schertzer [164] in which nonlocal interactions[2] are treated separately and included analytically in the kinetic-energy transfer term in the EDQNM spectral evolution equation. Details are also given in Lesieur ([170], pp. 231–235). Wavenumbers are discretized logarithmically in the form

$$k_L = \delta k\ 2^{(L-1)/F}, \qquad (4.9)$$

with L ranging from 1 to a maximum value L_S. In all calculations, F was taken equal to 8, which is twice as large as used in former calculations of this type done in Grenoble and should guarantee a higher precision.[3] Calling k_{max} the maximum wavenumber, we have also

$$\frac{k_{max}}{k_i(0)} = A R_{k_i(0)}^{3/4}, \qquad (4.10)$$

---

[2] Nonlocal interactions are those involving extremely distinct wavenumbers and thus very elongated triads.

[3] Comparisons with calculations done with $F = 4$ show that the difference of results is not very substantial, and so the latter value should be recommended, considering the much shorter computational times in this case.

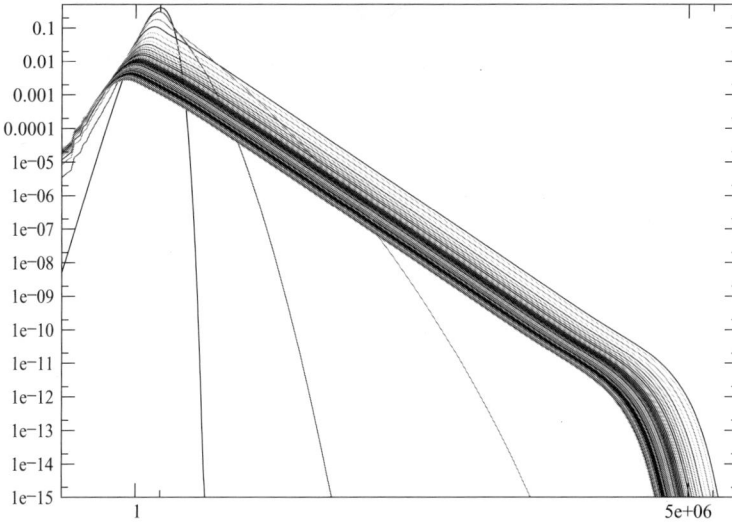

Figure 4.1. Kinetic-energy spectrum evolution in a decaying EDQNM calculation with $R_{k_i(0)} \approx 1.70 \times 10^9$.

with A equal to 1 and 3 in the decaying and forced calculations, respectively ($R_{k_i(0)}$ is a large-scale Reynolds number defined momentarily). This is lower than the value 8 proposed in Lesieur [170], but it permits a good-enough capturing of the dissipative range and results in a substantial reduction of computing time. These calculations have in fact been done on a PC/LINUX machine.

In decaying calculations, the initial kinetic-energy spectrum is

$$ E(k, 0) = A_s \, k^s \, \exp\left[-\frac{s}{2} \frac{k^2}{k_i(0)^2} \right], \qquad (4.11) $$

where A_s is a normalization constant chosen such that $\int_0^{k_{\max}} E(k, 0)dk = \frac{1}{2}v_0^2 = \frac{1}{2}$. The time unit is the initial large-eddy turnover time $[v_0 k_i(0)]^{-1}$. The constant a_1 corresponds to $C_K = 1.40$. The initial large-scale Reynolds number is $R_{k_i(0)} = v_0/\nu k_i(0)$.

We first present a calculation with $s = 8$, $\delta k = 0.125$, $k_i(0) = 2$, and $R_{k_i(0)} \approx 1.70 \times 10^9$. Figure 4.1 displays the time evolution of the kinetic-energy spectrum $E(k, t)$ for this run, up to 100 turnover times. We see very clearly the establishment of an ultraviolet inertial-type range whose slope may be checked to be (on this log–log plot) very close to the $k^{-5/3}$ Kolmogorov law along more than five decades. In fact this point will be explored later by considering compensated spectra $\epsilon^{-2/3} k^{5/3} E(k, t)$. We see also on the figure the rapid formation of a k^4 infrared spectrum. This corresponds to the k^4 infrared spectral backscatter, which will be discussed later. At the end of the evolution ($t = 100$), the Reynolds number based on the Taylor microscale and already defined in Chapter 1 is $R_\lambda \approx 72,600$. This is huge compared with laboratory

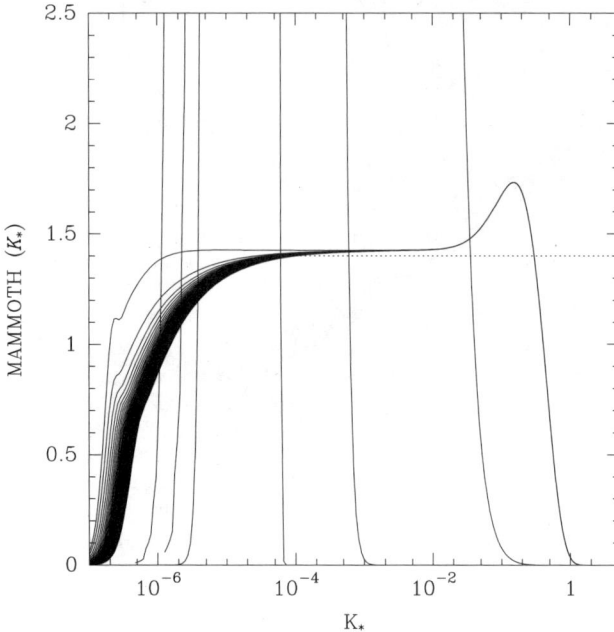

Figure 4.2. Kolmogorov-compensated kinetic-energy spectrum evolution (in dissipative units) corresponding to the EDQNM run of Figure 4.1.

or even environmental situations and might be encountered in astrophysics. Figure 4.2 shows for the run the Mammoth-shaped function $M(k_*, t)$ (introduced Chapter 1 on the r.h.s. of Eq. (1.47), with $k_* = k\eta$). The vertical lines correspond to spectra early at times. At later times, we get a perfect superposition of the curves at high wavenumbers, which indicates the validity of Kolmogorov similarity. At low wavenumbers, the dark area represents a decay of compensated spectra, which can be interpreted as the "Mammoth losing fat from the back."[4] At the end of the evolution there is a two-decade real compensated plateau at $C_K = 1.4$, and the spectral-bump size is one decade long. It is clear here that the limit of infinite Reynolds number, which would yield a Kolmogorov $k^{-5/3}$ spectrum extending to infinity, is just a mathematical view that cannot be reproduced in these calculations. However, Lesieur and Ossia [174] show that at such a high Reynolds number a limit curve is obtained for the skewness $s(t)$ defined by Eq. (1.33). The curve can be obtained from the following relation (see Orszag [225]):

$$s(t) = \left(\frac{135}{98}\right)^{1/2} D(t)^{-3/2} \int_0^{+\infty} k^2 T(k, t) dk, \qquad (4.12)$$

---

[4] A former French minister for education used to say that he would remove the fat off the national education mammoth.

where $T(k, t)$ is the kinetic-energy transfer given here by the r.h.s. of Eq. (4.5). The time evolution from zero to infinity of this limit skewness displays first a rise to the maximum value of 1.132 attained at $t \approx 4.1$, then an abrupt drop to a plateau value of 0.547 reached at $t \approx 4.8$, and is conserved exactly above up to $t = 100$. This evolution is explained in Lesieur [170] as a transition between an initial inviscid skewness growth[5] to a skewness determined by a balance between vortex stretching and molecular dissipation terms in the r.h.s. of the enstrophy time-evolution equation. This yields a skewness constant with time if enstrophy and palinstrophy are assumed to be dominated by inertial and dissipative wavenumbers and scale on Kolmogorov dissipative units (Batchelor [18], Orszag [225]).

Let us return to the EDQNM Mammoth-shape compensated spectra. As stressed in Chapter 1, similar behaviors may be obtained from experimental data, with similar type of scalings, as reviewed for instance by Coantic and Lasserre [47], who have developed an analytical model to account satisfactorily for Reynolds-number changes in the experiments. The bump-shaped spectrum had already been observed in the EDQNM calculations of André and Lesieur [6]. The bump was interpreted as a "bottleneck effect" by Falkovich [89]. We will return to this point later. Concerning the departure from Kolmogorov similarity at small wavenumbers, we will see that the latter cannot be achieved with the $s = 8$ value taken initially; it is only for $s = 1$ that it may hold.

4.3 EDQNM plateau-peak model

As we did for the deterministic velocity and scalar fluctuations, we split the EDQNM kinetic-energy and scalar-variance transfers into interactions involving only modes smaller than k_C and those involving the others. The equations for the supergrid-scale velocity $\bar{E}(k, t)$ and scalar $\bar{E}_\rho(k, t)$ spectra are, respectively,

$$\left(\frac{\partial}{\partial t} + 2\nu k^2\right) \bar{E}(k, t) = T_{<k_C}(k, t) + T_{>k_C}(k, t) \qquad (4.13)$$

and

$$\left(\frac{\partial}{\partial t} + 2\kappa k^2\right) \bar{E}_\rho(k, t) = T^\rho_{<k_C}(k, t) + T^\rho_{>k_C}(k, t), \qquad (4.14)$$

where $T_{<k_C}(k, t)$ and $T^\rho_{<k_C}(k, t)$ are the spectral transfers corresponding to resolved triads such that $k, p, q \leq k_C$ and $T_{>k_C}$ (resp. $T^\rho_{>k_C}$) transfer to modes

---

[5] We recall that Lesieur ([170], pp. 190–191) has shown for an initial-value problem in the framework of the Euler equation that, if $s(t)$ grows with time, or remains constant, or even decays slower than t^{-1}, then enstrophy will blow up in a finite time.

such that $k < k_C$, p and (or) $q > k_C$. We assume first that $k \ll k_C$ with both modes being larger than k_i, the kinetic-energy peak. Expansions in powers of the small parameter k/k_C yield to the lowest order

$$T_{>k_C}(k, t) = -2\nu_t^\infty \, k^2 \, \bar{E}(k, t), \tag{4.15}$$

$$\nu_t^\infty = \frac{1}{15} \int_{k_C}^\infty \theta_{0pp} \left[5E(p, t) + p \frac{\partial E(p, t)}{\partial p} \right] dp, \tag{4.16}$$

$$T_{>k_C}^\rho(k, t) = -2\kappa_t^\infty \, k^2 \, \bar{E}_T(k, t), \tag{4.17}$$

$$\kappa_t^\infty = \frac{2}{3} \int_{k_C}^\infty \theta_{0pp}^\rho \, E(p, t) \, dp. \tag{4.18}$$

Let us start by assuming a $k^{-5/3}$ inertial range at wavenumbers greater than k_C. We obtain

$$\nu_t^\infty = 0.441 \, C_K^{-3/2} \left[\frac{E(k_C)}{k_C} \right]^{1/2} \tag{4.19}$$

and

$$\kappa_t^\infty = \frac{\nu_t^\infty}{Pr^{(t)}} \tag{4.20}$$

with

$$Pr^{(t)} = 0.6. \tag{4.21}$$

Here, $E(k_C)$ is the kinetic-energy spectrum at the cutoff k_C. The 0.6 value for the Prandtl number is in fact the highest one permitted by the choice of two further adjustable constants arising in the EDQNM passive-scalar equation (see [170]). If we assume for instance a Kolmogorov constant of 1.4 in the energy cascade, the constant in front of Eq. (4.19) will be 0.267. When k is close to k_C, the numerical evaluation of the EDQNM transfers yields

$$T_{>k_C}(k, t) = -2\nu_t(k|k_C) \, k^2 \, \bar{E}(k, t) \tag{4.22}$$

and

$$T_{>k_C}^\rho(k, t) = -2\kappa_t(k|k_C) \, k^2 \, \bar{E}(k, t) \tag{4.23}$$

with

$$\nu_t(k|k_C) = K \left(\frac{k}{k_C} \right) \nu_t^\infty \tag{4.24}$$

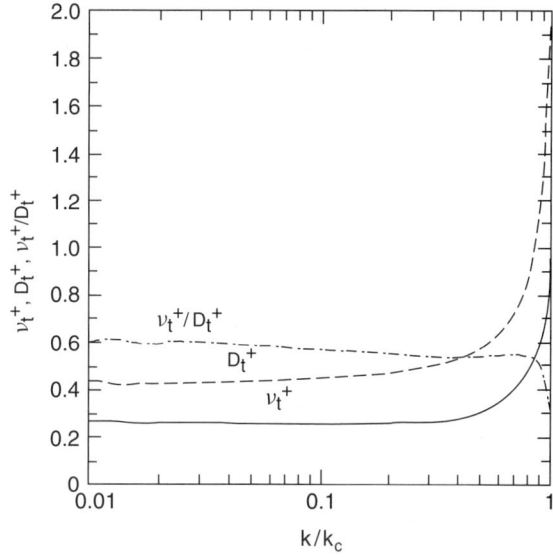

Figure 4.3. Eddy viscosity, eddy diffusivity, and turbulent Prandtl number in spectral space determined using the EDQNM theory. (From [42].)

and

$$\kappa_t(k|k_C) = C\left(\frac{k}{k_C}\right)\kappa_t^{\infty}, \qquad (4.25)$$

where ν_t^{∞} and κ_t^{∞} are the asymptotic values given by Eqs (4.19), (4.20), and (4.21), and $K(x)$ and $C(x)$ are nondimensional functions equal to 1 for $x = 0$. As shown also by Kraichnan's test-field model calculations [147], $K(x)$ has a plateau value at 1 up to $k/k_C \approx 1/3$. Above, it displays a strong peak (cusp behavior). Let us mention that Kraichnan did not point out the scaling of the eddy viscosity against $[E(k_C)/k_C]^{1/2}$, which turns out to be essential for LES purposes. Indeed, when the energy spectrum decreases rapidly at infinity (for instance during the initial stage of decay in isotropic turbulence), the eddy viscosity will be very low and inactive. However, we have $[E(k_C)/k_C]^{1/2} \sim \epsilon^{1/3}k_C^{-4/3}$ in an inertial-range expression. If we keep this inertial-range-type eddy viscosity before the establishment of the $k^{-5/3}$ range and evaluate ϵ as proportional to $E_c^{3/2}k_i$, it may substantially increase the eddy viscosity and work against the cascade development. We will explain in the following that the plateau-peak model may be generalized to spectra different from the Kolmogorov one at the cutoff (spectral-dynamic model).

It was shown in [42] that $C(x)$ behaves qualitatively as $K(x)$ (plateau at 1 and positive peak) and that the spectral turbulent Prandtl number $\nu_t(k|k_C)/\kappa_t(k|k_C)$ is approximately constant and thus equal to 0.6 as given by Eq. (4.21). These three quantities (eddy viscosity, eddy diffusivity, and turbulent Prandtl number) taken from [42] are shown in Figure 4.3 as a function of k/k_C. In the figure, the eddy coefficients are normalized by $\sqrt{E(k_C)/k_C}$ with $C_K = 1.4$.

It is clear that the plateau part corresponds to the usual eddy-coefficients assumption when one goes back to physical space,[6] and thus the "peak" part goes beyond the scale-separation assumption inherent in the classical eddy-viscosity and diffusivity concepts. The peak is mostly due to semilocal interactions across k_C: Near the cutoff wavenumber, the main nonlinear interactions between the resolved and unresolved scales involve the smallest eddies of the former and the largest eddies of the latter (such that $p \ll k \sim q \sim k_C$). The peak also contains possible backscatter contributions (which are however very small if k_C lies in a Kolmogorov cascade) coming from subgrid modes larger than k_C. This point will be detailed in the following.

As shown in [43], the plateau-peak behavior of $K(x)$ can be approximately expresssed with the following analytical expression:

$$K(x) = 1 + 34.5\, e^{-3.03/x}. \tag{4.26}$$

We will see later another analytic expression of this spectral eddy viscosity in terms of hyperviscosities.

The plateau-peak model consists of using these eddy viscosities in the deterministic equations (4.1) and (4.3). One advantage of such a subgrid-scale modeling is that it is correct from an energy-transfer viewpoint. It is also able to deal with a continuous spectrum at the cutoff, which is a great asset with respect to the plain eddy-viscosity assumption in physical space. However, the assumption of real eddy coefficients is constraining and neglects the possible phase effects arising in the neighborhood of k_C.

4.3.1 Spectral-dynamic model

Another drawback of the plateau-peak model is that it is restricted to the case in which k_C lies within a $k^{-5/3}$ Kolmogorov cascade. Fortunately, this can be cured by introducing the spectral-dynamic model. We assume now that the kinetic-energy spectrum is $\propto k^{-m}$ for $k > k_C$ with m not necessarily equal to 5/3. We modify the spectral eddy viscosity as

$$\nu_t(k|k_C) = 0.31\, C_K^{-3/2} \sqrt{3-m}\, \frac{5-m}{m+1} K\left(\frac{k}{k_C}\right) \left[\frac{E(k_C)}{k_C}\right]^{1/2} \tag{4.27}$$

for $m \leq 3$. This expression is exact for $k \ll k_C$ within the same nonlocal expansions of the EDQNM theory, as shown in Métais and Lesieur [205]. We retain the peak shape through $K(k/k_C)$ to be consistent with the Kolmogorov spectrum expression of the eddy viscosity. For $m > 3$, the scaling is no longer

---

[6] There is, however, a slight difference at this level because going back to physical space will give ν_t^∞ multiplied by the filtered-velocity Laplacian, whereas, in the physical-space formalism, the eddy viscosity is under a divergence operator in Eq. (3.19).

valid, and the eddy viscosity will be set equal to zero. Indeed, we are very close to a DNS for such spectra. In the spectral-dynamic model, the exponent m is determined through the LES with the aid of least-squares fits of the kinetic-energy spectrum close to the cutoff. We may also check that the turbulent Prandtl number is given by

$$Pr^t = 0.18\,(5 - m) \tag{4.28}$$

(see Métais and Lesieur [205] and Lesieur [170], p. 386). This value does not depend of the Kolmogorov and model constants. Being able to use a variable turbulent Prandtl number is a great advantage in LES of heated or variable-density flows. This possibility exists also for the dynamic models in physical space such as the dynamic Smagorinsky model presented in Chapter 3.

4.3.2 Spectral random backscatter

There are many discussions on LES related to the concept of random backscatter, one aspect of which in physical space is the negativeness of the eddy viscosity in local regions. We give here some elements of this discussion in Fourier space. We return to the EDQNM kinetic-energy transfer $T(k, t)$ in three-dimensional isotropic turbulence given by the r.h.s. of Eq. (4.5). Such a transfer may be rewritten by a symmetrization with respect to p and q in the integrand: In the first term $a(k, p, q) = (1/2)[b(k, p, q) + b(k, q, p)]$ appears, which may be shown to be positive (see Orszag [225] and Lesieur [170]). The second term is proportional to $k^2 E(k, t)$. This ensures the realizability (positiveness of the kinetic-energy spectrum) of the closure. We consider now some arbitrary cutoff wavenumber k_C, which is not necessarily in the middle of an inertial range. The subgrid kinetic-energy transfer across k_C is then

$$T_{sg}(k) = A_{BS} - B_D, \tag{4.29}$$

where A_{BS}, the backscatter term, is given by

$$A_{BS} = k^4 \int_{k_C}^{\infty} dp \int_{k/2p}^{1} \frac{1 - z^2}{q^2}$$
$$\times \left[1 + \frac{p^2}{q^2} + \left(\frac{k}{q} - 2\frac{p}{q}z \right)^2 \right] \theta_{kpq} E(p)E(q)\,dz. \tag{4.30}$$

This term is obviously positive. The second term can be written as

$$B_D = k^2 E(k) \int_{k_C}^{\infty} dp \int_{k/2p}^{1} \theta_{kpq}(1 - z^2)$$
$$\times \left[\left(\frac{p^2}{q^2} - \frac{pz}{k} \right) \frac{p^2}{q^2} E(q) + \left(1 - \frac{p^2}{q^2} + \frac{p^3 z}{kq^2} \right) E(p) \right] dz. \tag{4.31}$$

These expressions can be simplified if $k \ll k_C$ (which implies that p and q are of the same order). Then,

$$A_{BS} = \frac{14}{15}k^4 \int_{k_C}^{\infty} \theta_{0pp} \frac{E(p)^2}{p^2} dp, \qquad (4.32)$$

$$B_D = 2v_t^{\infty} k^2 E(k), \qquad (4.33)$$

where the eddy viscosity v_t^{∞} has been given in Eq. (4.16). If k and k_C both lie in the inertial range (with $k \ll k_C$), the k^4 backscatter is of the order of $k^4 \theta_{0,k_C,k_C} k_C^{-1} E(k_C)^2$, and the eddy-viscosity contribution is of the order of $k^2 E(k) \theta_{0,k_C,k_C} k_C E(k_C)$. Hence, in this case

$$\frac{A_{BS}}{B_D} \sim \left(\frac{k}{k_C}\right)^2 \frac{E(k_C)}{E(k)}, \qquad (4.34)$$

which is very small for any decreasing kinetic-energy spectrum. This justifies the fact that the plateau part of the spectral eddy viscosity considered here does not include any k^4 backscatter contribution. However, backscatter is important when k is close to k_C, but the coefficient in front of k^4 is not a simple function of k; moreover, it is difficult to tell the exact k dependence of the backscatter in this case or of the eddy-viscosity term B_D. What is certain is that the plateau-peak eddy viscosity does properly include the backscatter at the level of correct kinetic-energy exchanges.

In fact, the k^4 backscatter transfer plays an important role in the infrared part of the spectrum ($k \to 0$). We assume that $k_C = k_i$ corresponds to the peak of the spectrum and again $k \ll k_C$. Now the backscatter given by Eq. (4.32) dominates the local transfers. It injects energy in very large scales through resonant interaction of two energetic modes, and it is responsible for the immediate emergence of an infrared k^4 spectrum in isotropic decaying turbulence when energy is injected initially at a peak at k_i. This point, predicted by two-point closures (see Lesieur and Schertzer [164] and Lesieur [170]), was first checked in LES of isotropic turbulence by Lesieur and Rogallo ([165]; see Figure 4.4) using the plateau-peak model, and we will confirm it with LES using the spectral-dynamic model.

In forced stationary turbulence obtained when a random statistically stationary forcing is applied on a narrow spectral band around k_i, the net infrared transfer is given by the combination of the backscatter and the eddy-viscous drain. It should vanish because the energy spectrum is time invariant. There is then a balance between the k^4 backscatter and the $k^2 E(k)$ drain, which yields a k^2 equipartition spectrum.

We stress finally that in a turbulent mixing-layer calculation, Leith [162] used a k^4 random backscatter forcing as a way to inject energy into the large scales.

4.4 Return to the bump

We have already mentioned for decaying isotropic turbulence the "bump" existing at the edge of the Kolmogorov $k^{-5/3}$ inertial range before the dissipative range. In fact, forced EDQNM calculations with a narrow forcing at k_i do show the persistence of the bump (Mestayer et al. [203], Lesieur and Ossia [174]). In the calculations of Mestayer et al. [203], the bump did disappear with the removal of nonlocal triads (k, p, q) of the type $k < ap$ (with $a \approx 2^{1/F} - 1 \approx 0.2$ when taking $F = 4$). These elongated nonlocal interactions correspond to an energy flux given by (see [170], p. 233, for details)

$$
\Pi_{\text{EL}}(k, t) = \frac{2}{15} \int_0^k k'^2 E(k')dk' \int_{\sup(k,k'/a)}^\infty \theta_{k'pp} [5E(p) + p\frac{\partial E}{\partial p}]dp
$$

$$
- \frac{14}{15} \int_0^k k'^4 dk' \int_{\sup(k,k'/a)}^\infty \theta_{k'pp} \frac{E(p)^2}{p^2}dp. \tag{4.35}
$$

The first term in Eq. (4.35) is of the "eddy-viscous type"; the second is of the "backscatter type," but the latter may be checked to be negligible in the energy cascade, as already stressed. No real explanation for the bump disappearance is given in Mestayer et al. [203], who just note that "the bumps appear to result mainly from a lack of erosion of the spectra by elongated non-local interactions when approaching the viscous cutoff." Falkovich [89] interpreted the bump as a "bottleneck phenomenon . . . where a viscous suppression of small scale modes removes some triads from nonlinear interactions . . . which leads to a pileup of the energy in the inertial interval of scales." In fact, this may be made more quantitative by looking back at the evaluation of the elongated nonlocal flux given by Eq. (4.35) carried out in [203]. It is positive and approximately constant in the inertial range. We will assume that it would remain constant in a $k^{-5/3}$ range extending to infinity. However, because of its structure in terms of integrals to infinity upon the energy spectrum, the elongated flux should start to decrease rapidly when feeling nonlocally the dissipative range, which is much further upstream. If we assume that the local and other nonlocal fluxes are not yet affected by dissipation, and hence are still constant, the global flux will be decreased, implying a positive kinetic-energy transfer, resulting in the bump.

4.5 Other types of spectral eddy viscosities

4.5.1 Heisenberg's eddy viscosity

In fact, the concept of a wavenumber-dependant eddy viscosity may already be found in Heisenberg ([120], see also Mc Comb [200] for details). Heisenberg introduced this eddy viscosity to model the evolution of the kinetic-energy spectrum. Within this model, and as recalled by Schumann [260], the derivative of the eddy viscosity with respect to k is proportional to $-\sqrt{E(k)/k^3}$. If we assume some power-law dependence for the kinetic-energy spectrum, Heisenberg's eddy viscosity will indeed scale as $\sqrt{E(k)/k}$. This is a type of local spectral eddy viscosity, which is less rich than the nonlocal plateau-peak formulation. It was used by Aubry et al. [10] to model equivalent subgrid scales in the dynamical system describing the evolution of a turbulent boundary layer within a proper orthogonal decomposition (POD) approach. We recall that in the POD (see Holmes et al. [126] for a review), the velocity vector is projected on the eigenvectors of the Reynolds-stress tensor. In this context, ejection or sweep events occurring in the boundary layer appeared as particular events in a chaotic dynamical system.

4.5.2 RNG analysis

Another approach, the renormalization group (RNG) method, originally developed by Forster et al. [97] and Fournier [99] for isotropic turbulence, has been applied by Yakhot and Orszag [293] and McComb [200] to LES with an eddy viscosity proportional to $\sqrt{E(k_C)/k_C}$. Let us recall briefly the RNG formalism in Fournier's work. In classical RNG analysis applied to the physics of critical phenomena, the dimension d of space is considered as a variable parameter. In general, the problem can be solved analytically for the dimension $d = 4$. Then the solution for $d = 4 - \epsilon$ is obtained from this solution through expansions in powers of the parameter ϵ, which is assumed to be small. The solution for $d = 3$ is recovered by making $\epsilon = 1$. Although slightly awkward, the procedure works remarkably well for various problems such as spin dynamics in ferromagnetic systems. Forster et al. [97] adapted the method to the Navier–Stokes equation with a varying dimension of space. In contrast, Fournier works with a fixed dimension of space (three), and he considers a kinetic-energy forcing term proportional to k^{-r} with a varying exponent r. One supposes at a given time that energy is distributed on a wavenumber interval $[0, \Lambda]$. Let $\delta\Lambda \ll \Lambda$, and let $\Lambda - \delta\Lambda$ be a sort of cutoff wavenumber with $\delta\Lambda/\Lambda$ fixed. The velocities corresponding to modes in the shell $[\Lambda - \delta\Lambda, \Lambda]$ are solved, through Feynman diagrammatic perturbation techniques involving Green-function operators, in terms of modes smaller than $\Lambda - \delta\Lambda$. Statistical independence between the "subgrid" and "supergrid" modes is further assumed. A new Navier–Stokes equation with renormalized eddy viscosity and forcing, involving the wavenumber interval $[0, \Lambda - \delta\Lambda]$, is written. One has thus eliminated ("decimated") the shell $[\Lambda - \delta\Lambda, \Lambda]$. As stressed by Lesieur ([170] p. 253), other terms, called "nonpertinent," still arise at this level, but these will vanish after an infinite number of decimations. Then the operation is iterated an infinite number of times to let the cutoff Λ go to zero. The small parameter here is $\epsilon = 3 + r$. For $\epsilon > 0$, one obtains (Fournier and Frisch [100], Lesieur [170], p. 255) an eddy viscosity proportional to $\sqrt{E(\Lambda)/\Lambda}$. However, to obtain a Kolmogorov $k^{-5/3}$ kinetic-energy spectrum requires $r = 1$, so that the "small" parameter is now 4, which is excessive and cannot guarantee the convergence of the expansions. Furthermore, this expression of the renormalized eddy viscositiy is valid only for $\Lambda \to 0$, whereas it is used for LES purposes with a finite cutoff for which the nonpertinent terms cannot be neglected. Finally, there is no general consensus about the determination of the numerical constant arising in the eddy viscosity.

These results indicate that the plateau-peak model has the richest dynamics of all the Heisenberg-type $\sim\sqrt{E(k_C)/k_C}$ eddy viscosities.

4.6 Anterior spectral LES of isotropic turbulence

The plateau-peak eddy viscosity was applied by Chollet and Lesieur [41] to the first spectral LES of three-dimensional isotropic turbulence (a pseudo-spectral method with a resolution of 32^3 Fourier collocation modes). They studied decaying turbulence; there is no molecular viscosity,[7] and the initial energy spectrum decreases rapidly at infinity. During the first stage the kinetic energy is transferred toward k_C accompanied by a growth of the resolved enstrophy $D(t)$. At about four initial large-eddy turnover times $D(0)^{-1/2}$, the enstrophy reaches a maximum and decreases, whereas the kinetic-energy spectrum decays self-similarly with an approximate $k^{-5/3}$ slope.

Large-eddy simulations of a passive scalar at the same resolution were performed by these authors in 1982 with qualitatively the same results and the formation of a $k^{-5/3}$ Corrsin–Oboukhov inertial-convective scalar spectrum. These results are presented in Lesieur ([170], p. 389). However, using 32^3 collocation points gives extremely low resolution and is totally unable to capture the fine features of isotropic turbulence. We show in Figure 4.4 an analogous LES at a resolution of 128^3 Fourier modes[8] carried out by Lesieur and Rogallo [165]. The initial velocity and scalar spectra are proportional with a Gaussian ultraviolet behavior and a k^8 infrared spectrum. It can be checked that Kolmogorov and Corrsin–Oboukhov $k^{-5/3}$ cascades are established. Afterward, the kinetic-energy spectrum decays self-similarly with a spectral slope between $-5/3$ and -2. The scalar spectrum seems to have a

---

[7] It is a nice behavior of these large-eddy simulations to allow for "Euler LES" without any numerical energy diffusion. The question posed is of course of the relevance of the solutions found with respect to real solutions of Euler equations or of the Navier–Stokes equation in the limit of zero viscosity.

[8] Such a simulation was not dealiased, but it is now well recognized that, in contrast to DNS, aliasing effects may be important in spectral LES and should be eliminated.

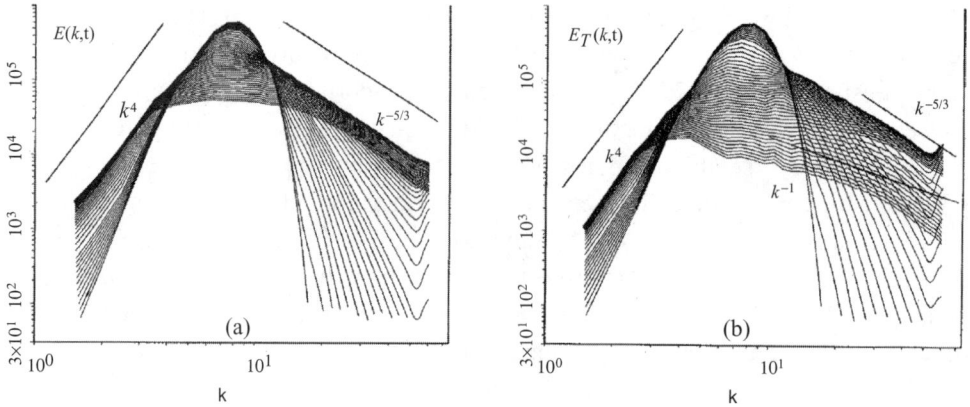

Figure 4.4. Three-dimensional isotropic decaying turbulence showing decay of kinetic-energy (a) and passive-scalar (b) spectra in the LES of Lesieur and Rogallo [165] using the plateau-peak eddy viscosity.

very short inertial-convective range close to the cutoff and a very wide range shallower than k^{-1} in the large scales. Here, the scalar decays in time much faster than the temperature. This anomalous range was explained by Métais and Lesieur [205] as due to the quasi-two-dimensional character of the scalar diffusion in the large scales, leading to large-scale intermittency of the scalar. More precisely, the scalar diffusion seems to be dominated by the effect of the coherent vortices already considered in Chapter 2. More details on this anomalous k^{-1} range may be found in Lesieur ([170], p. 211).

4.6.1 Double filtering in Fourier space

These spectral LESs of decaying isotropic turbulence and associated scalar mixing, together with those of Métais and Lesieur [205], have been used to compute directly the spectral eddy viscosity and diffusivity. The method is the same as that employed by Domaradzki et al. [70] for a DNS: One defines a fictitious cutoff wavenumber $k'_C = k_C/2$ across which the kinetic-energy transfer T and scalar transfer T^ρ are evaluated. Because we deal with a LES, the latter corresponds to triadic interactions such that $k < k'_C$, p and (or) $q > k'_C$ and $p, q < k_C$. These are termed $T_{>k'_C}^{<k_C}(k, t)$ and $T_{>k'_C}^{T<k_C}(k, t)$. They correspond to resolved transfers and satisfy energetic equalities of the type

$$T_{>k'_C}^{<k_C}(k, t) = T_{>k'_C}(k, t) - T_{>k_C}(k, t), \tag{4.36}$$

where $T_{>k'_C}$ and $T_{>k_C}$ are the total kinetic energy transfers across k'_C and k_C. It is important to note that Eq. (4.36) is the exact energetic equivalent in spectral space of Germano's identity if one works in Fourier space with sharp filters. A similar relation holds for $T_{k'_C}^{T<k_C}$. Dividing these equations by $-2k^2\, E(k, t)$ and $-2k^2\, E_\rho(k, t)$ gives the resolved spectral eddy-viscosity and diffusivity.

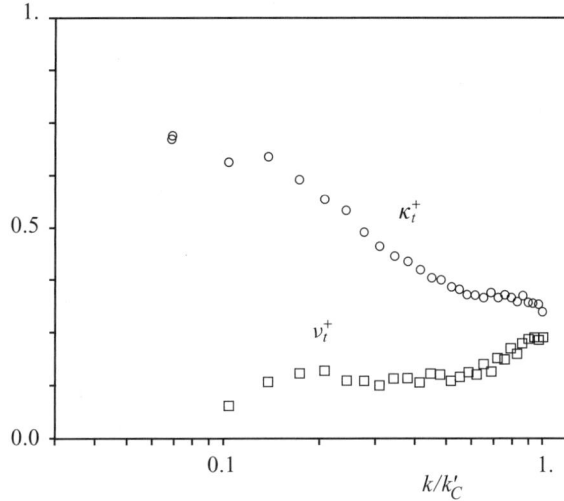

Figure 4.5. Resolved eddy viscosity and diffusivity evaluated through a double filtering in LES of isotropic decaying turbulence. (From Métais and Lesieur [205].)

Figure 4.5 shows these functions normalized by $[E(k'_{\rm C})/k'_{\rm C}]^{1/2}$. Similar results have been found in Lesieur and Rogallo [165]. The figure demonstrates that the plateau-peak behavior does exist for the eddy viscosity but is questionable for the eddy diffusivity. This anomaly is obviously related to the anomalous scalar k^{-1} range previously mentioned. It was stressed by Lesieur ([170], p. 392) that this anomalous scalar range still exists in a DNS of decaying isotropic turbulence: In this case, the double filtering yields a plateau-peak eddy viscosity with a plateau value of approximately zero, as was discovered by Domaradzki et al. [70]. The eddy diffusivity, in contrast, still behaves as in the LES. In fact, Métais and Lesieur [205] have checked that the anomaly disappears when the temperature is no longer passive and is coupled with the velocity within the framework of the Boussinesq approximation (stable stratification). It is possible that the same holds for compressible turbulence, which would legitimize the use of the plateau-peak eddy diffusivity in this case.

4.7 EDQNM infrared backscatter and self-similarity

We return now to the EDQNM analysis of Lesieur and Ossia [174] and show that it is only at $s = 1$ that the kinetic-energy spectrum may have a global self-similarity at entire scales from the energy-containing to the dissipative ones. The derivation is borrowed from Lesieur and Schertzer [164], who applied it to the EDQNM spectral equation. We present a generalization that does not require use of closure. The first point is to remark that such a global self-similarity necessarily implies that the integral and dissipative scales l and $l_{\rm D}$ are proportional, with their ratio being time independent. If a Karman–Howarth self-similarity is assumed, the kinetic-energy and transfer spectra are, respectively,

$$E(k, t) = v^2 l \, F(kl), \quad T(k, t) = v^3 \, T_1(kl), \quad (4.37)$$

where the functions F and T_1 are nondimensional. We assume in fact that a regime is reached such that all the quantities have an algebraic time dependance, and thus we can write

$$E(k, t) = t^n \, G(k'), \quad T(k, t) = t^{3(n-m)/2} \, T'(k') \quad (4.38)$$

with $k' = kt^m$, $v^2 \propto t^{n-m}$, and $l \propto t^m$. Notice here that G and T' are dimensional functions of the dimensional argument k'. Substituting these expressions into the kinetic-energy spectrum evolution equation

$$\frac{\partial E}{\partial t} + 2vk^2 E = T(k, t), \quad (4.39)$$

we obtain (after division by $t^{3(n-m)/2}$)

$$\left[nG + mk' \frac{dG}{dk'} \right] t^{(3m-n-2)/2} + 2vk'^2 G(k') t^{-(m+n)/2} = T'(k'). \quad (4.40)$$

In this equation, all the terms have to be time independent. We do have $3m - n - 2 = 0$ (a condition that we could have obtained by writing $\epsilon \sim v^3/l$) and $m + n = 0$, which finally implies that $m = 1/2, n = -1/2$, and α_E (such that $v^2 \propto t^{-\alpha_E}$) is equal to $m - n = 1$. It is a well-known result that such a global self-similarity, when applied to the infrared spectrum, implies a further condition. Indeed, relation (4.38) gives for an infrared kinetic-energy spectrum $\propto t^{\gamma_s} k^s$ (for which viscosity has a negligible effect if small enough)

$$\gamma_s = n + ms \quad (4.41)$$

and $s = 1 + 2\gamma_s$. We know that (in the EDQNM framework where a k^4 backscatter is assumed), γ_s is zero except for $s = 4$, where it is equal to 0.16 (see Lesieur and Schertzer [164] and Lesieur [170]), and the only possibility is thus $s = 1$. Hence, the large-scale Reynolds number R_l should be constant with time as well as $R_\lambda \sim \sqrt{R_l}$.

The question of permissible values for s is a controversial one. There are arguments in favor of $s = 4$ (see the review of Davidson [62]) and others in favor of $s = 2$ (Saffman [247]). Taking even values of s is compulsory if certain regularity conditions are fulfilled for the velocity-correlation tensor between two points when the distance goes to infinity. Mathematically, we may take initially odd values of s (such as 1 or 3) and even noninteger ones, as was proposed by Eyink and Thomson [88]. Working on the basis of an analogy with Burgers turbulence studied by Gurbatov et al. [116] with DNS, Eyink and Thomson propose the existence of a crossover dimension $s \approx 3.45$, above which a k^4 backscatter should appear. The crossover value is obtained

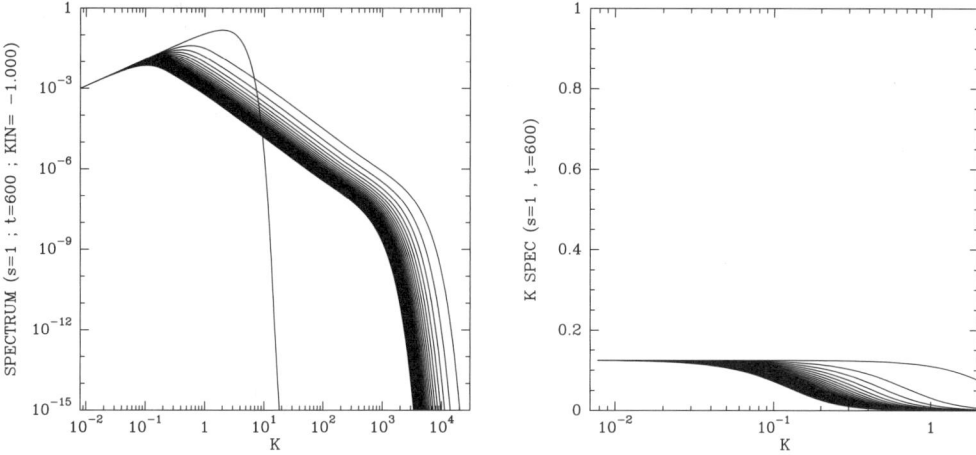

Figure 4.6. EDQNM decaying calculation for $s = 1$ showing time evolution of the kinetic-energy spectrum (left) and the spectrum multiplied by k^{-1} (right).

in the following way: They assume that kinetic energy decays with time like $t^{-1.38}$, the law obtained by Lesieur and Schertzer [164] for $s = 4$, and then equate this decay exponent to the exponent α_E, which is equal to

$$\alpha_E = \frac{2(s + 1)}{s + 3}, \tag{4.42}$$

an equation obtained assuming Karman–Howarth self-similar decaying spectra with permanence of large eddies.

We now present recent EDQNM decay calculations with an initial exponent s equal, respectively, to $1, 2, 3,$ and 4. Calculations are run up to 600 initial turnover times, with $k_i(0) = 2$, $\delta k = 2^{-7}$, and the initial large-scale Reynolds number $R_{k_i}(0) = 416{,}132$. Figure 4.6 (left) presents the kinetic-energy spectrum evolution (thirty-one curves) in the case $s = 1$. There is a good time invariance of infrared modes. This is confirmed by the compensated spectrum multiplied by k^{-1} (Figure 4.6 [right]), which has an invariant plateau at low k. We can check in this case (Lesieur and Ossia [174]) that kinetic energy decays like $t^{-1.00}$, which is in good agreement with the law of Eq. (4.42). Figure 4.7 (top left) shows the Mammoth function $M(k_*, t)$ for all the spectra of Figure 4.6 except the initial one. The thirty curves are perfectly superposed, which justifies for the EDQNM the results of the previous paragraph. The following calculations will confirm that it is only for this case that self-similarity at low and high wavenumbers holds. Let us look now at the time evolution of R_{k_i}, R_l, and R_λ. In Figure 4.8, we see the time constancy of the two latter Reynolds numbers when self-similarity is reached. However, the first displays oscillations. This shows that k_i is not strictly proportional to l and that the latter quantity is better for characterizing the large scales. In fact, this time constancy of the Reynolds number corresponds to a sort of nondissipative turbulence

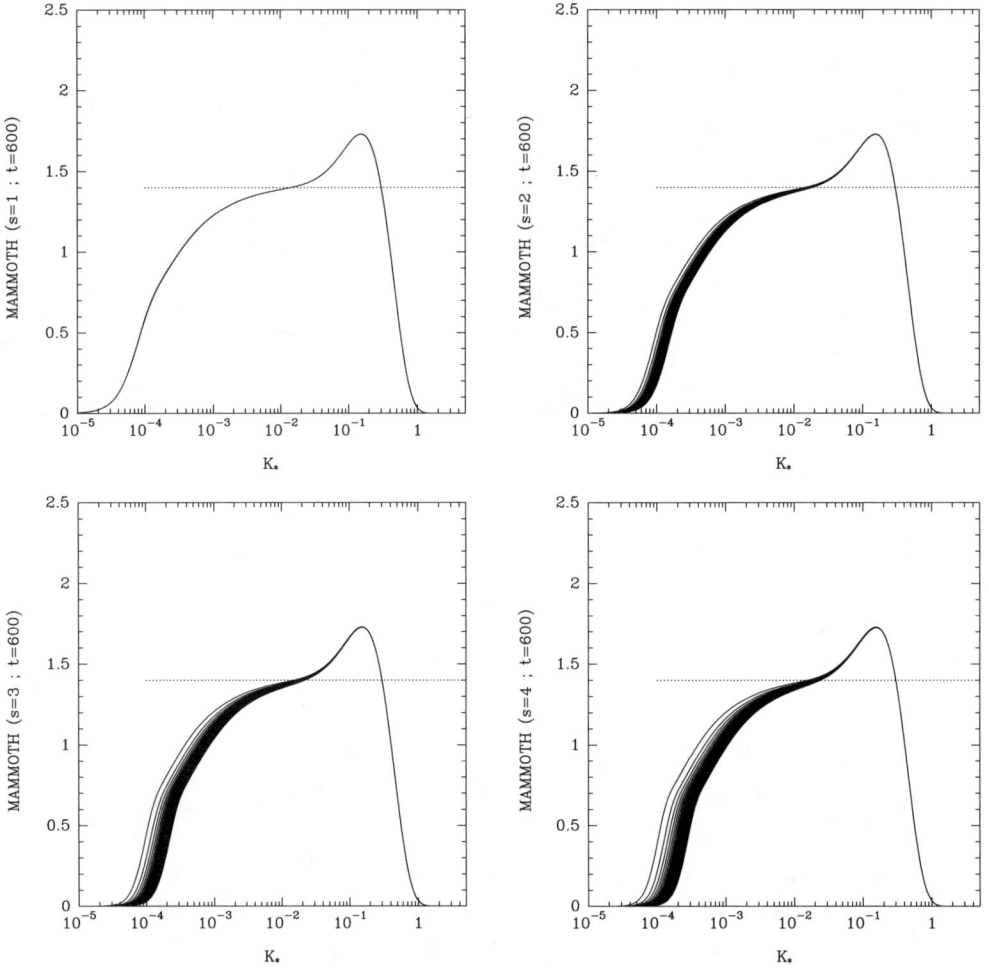

Figure 4.7. EDQNM decaying calculation for $s = 1$, $s = 2$, $s = 3$, and $s = 4$ showing time evolution of the Mammoth function $M(k_*, t)$.

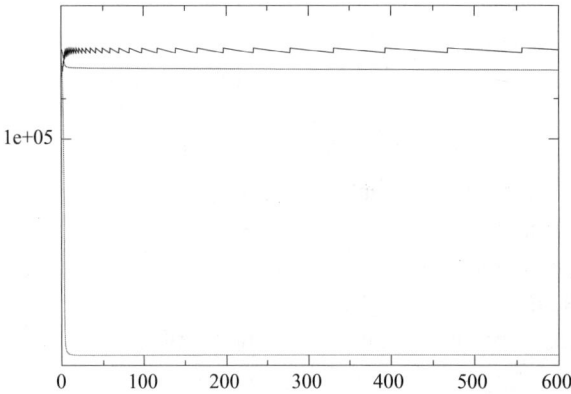

Figure 4.8. EDQNM decaying calculation for $s = 1$ showing time evolution in semilog coordinates of the Reynolds numbers (from top to bottom) R_{k_i}, R_l, R_λ.

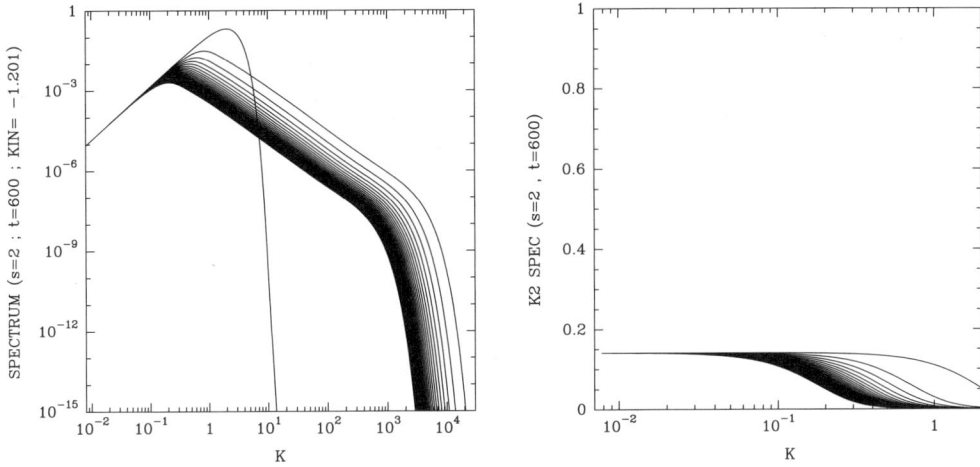

Figure 4.9. EDQNM decaying calculation for $s = 2$ showing time evolution of the kinetic-energy spectrum (left) and the spectrum multiplied by k^{-2} (right).

(because the Reynolds number does not vary with time), which is an exceptional (and not realistic[9]) situation. It should be noted also that the t^{-1} kinetic-energy decay law was proposed by Batchelor [18] but without stressing the necessary constraint $s = 1$ upon the infrared modes.

For other values of s larger than one, the Reynolds number will decay with time, as will be checked. We continue with $s = 2$, and spectra remain time invariant at low k (Figure 4.9). The kinetic energy decays at the end of the run as $t^{-1.201}$ [174]. This is very close to Saffman's 6/5 law [247], which can be recovered from Eq. (4.42) with $s = 2$. We will see later that this law is also very well verified in large-eddy simulations, which validates the EDQNM model as a tool for this type of problem. The Mammoth function has now lost self-similarity at low k (Figure 4.7). We still have quite a good time invariance at low k for $s = 3$, with, however, some weak backscatter. The corresponding curves are shown in Figures 4.10 and 4.7. The kinetic energy decays as $t^{-1.323}$, which is slightly above the 4/3 value obtained by replacing s by 3 in Eq. (4.42). This is certainly due to departure from self-similarity at low k. Indeed, it was checked in Lesieur and Ossia [174] that varying the Reynolds number for a given s has no effect on the decay law. The low-k time invariance is lost for $s = 4$ with a sensible k^4 backscatter, which is a feature particularly obvious on the compensated spectrum of Figure 4.11. Kinetic energy decays like $t^{-1.384}$, and the exponent has saturated. This is not far from Lesieur and Schertzer's 1.38 law. Figure 4.12 displays the Reynolds numbers evolution now with a global decay but maintaining an oscillatory behavior of R_{k_i}.

---

[9] We are not dreaming, however, that by some control of the large scales that would modify the infrared spectrum into a law proportional to k, turbulence might become in some sense eternal.

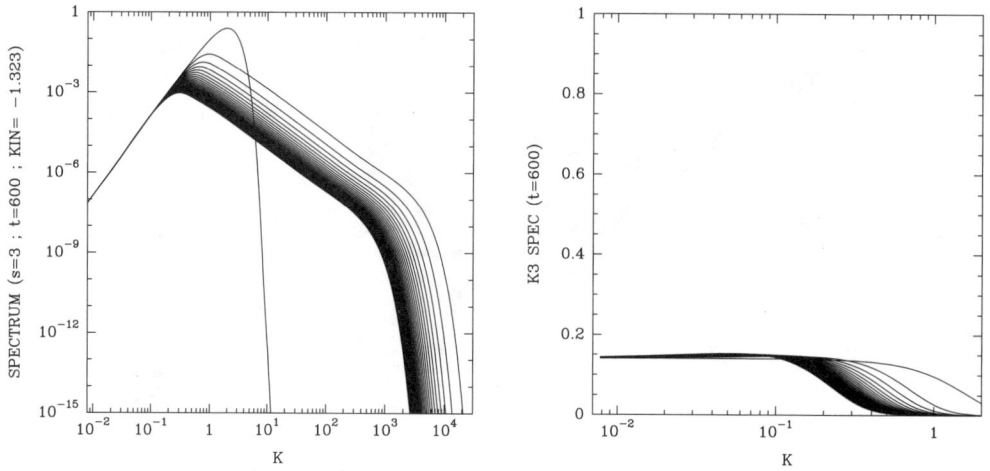

Figure 4.10. EDQNM decaying calculation for $s = 3$ showing time evolution of the kinetic-energy spectrum (left) and the spectrum multiplied by k^{-3} (right).

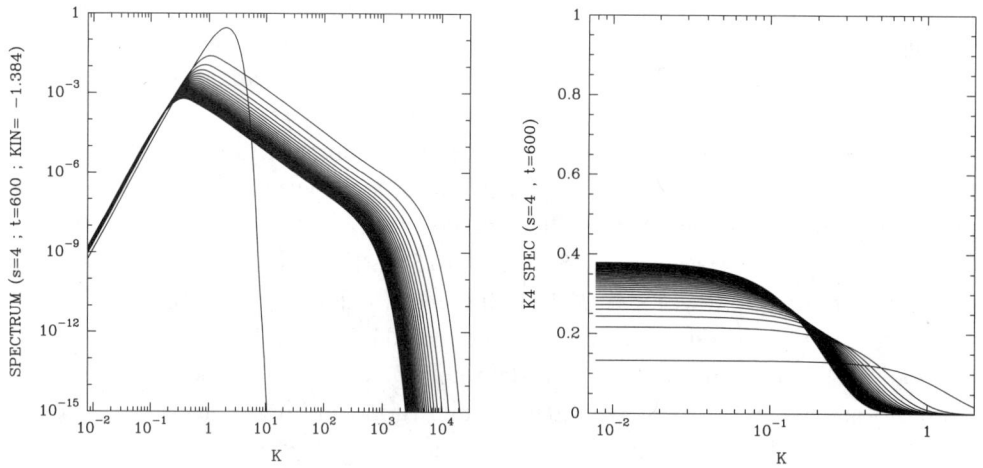

Figure 4.11. EDQNM decaying calculation for $s = 4$ showing time evolution of the kinetic-energy spectrum (left) and the spectrum multiplied by k^{-4} (right).

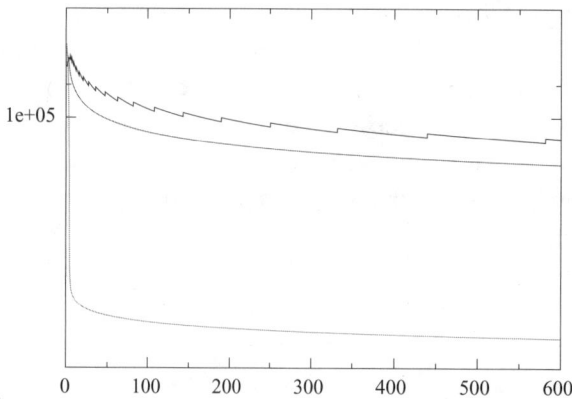

Figure 4.12. EDQNM decaying calculation for $s = 4$ showing time evolution in semilog coordinates of the Reynolds numbers (from top to bottom) R_{k_i}, R_l, R_λ.

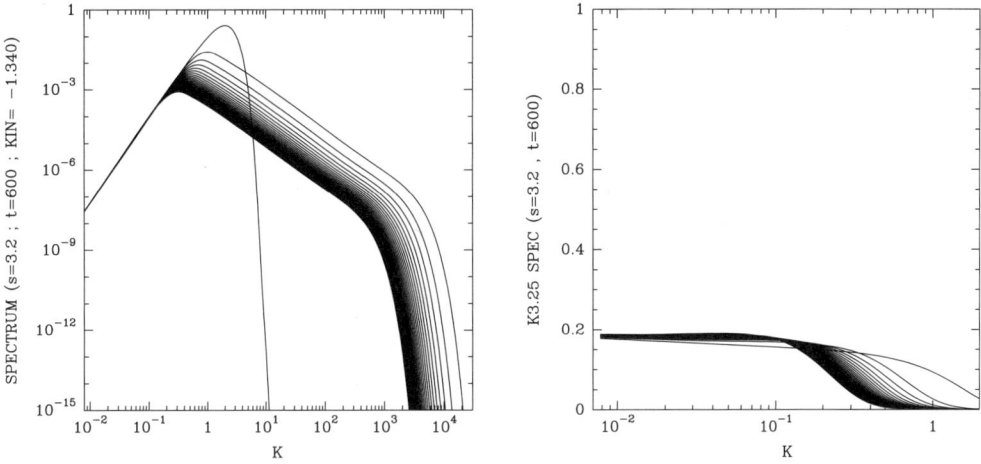

Figure 4.13. EDQNM decaying calculation for $s = 3.2$ showing time evolution of the kinetic-energy spectrum (left) and the spectrum multiplied by $k^{-3.25}$ (right).

We do not present results for s higher than 4, which can be found in Lesieur and Ossia [174], but concentrate now on the following noninteger values of s between 3 and 4. Here, the infrared spectrum will be compensated by the power of k that gives a flat plateau at $t = 600$. For $s = 3.2$ (Figure 4.13) a weak backscatter appears, which, from the infrared compensated spectrum, is best described by $k^{3.25}$. The kinetic-energy decay exponent saturates to 1.340, which is lower than the value 1.355 predicted by Eq. (4.42). For $s = 3.3$ (Figure 4.14), the backscatter remains weak and is now close to $k^{3.37}$. The kinetic-energy decay exponent saturates to 1.347 instead of the 1.365 prediction of Eq. (4.42). For $s = 3.4$ (Figure 4.15), backscatter becomes more

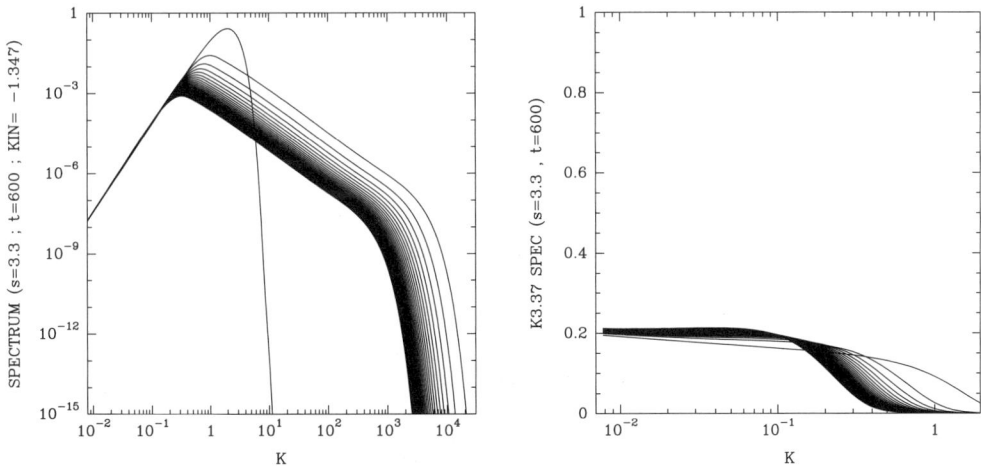

Figure 4.14. EDQNM decaying calculation for $s = 3.3$ showing time evolution of the kinetic-energy spectrum (left) and the spectrum multiplied by $k^{-3.37}$ (right).

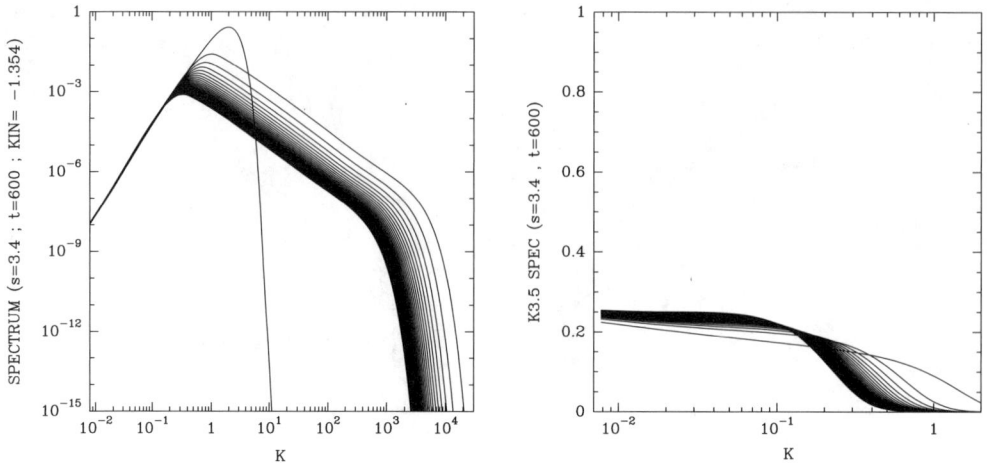

Figure 4.15. EDQNM decaying calculation for $s = 3.4$ showing time evolution of the kinetic-energy spectrum (left) and the spectrum multiplied by $k^{-3.5}$ (right).

important and is best described by $k^{3.5}$. We have $\alpha_E = 1.347$ instead of $11/8 = 1.375$ given by Eq. (4.42).

For $s = 3.5$ (Figure 4.16), backscatter slightly increases and is now described by $k^{3.6}$. We have $\alpha_E = 1.360$ instead of the analytical prediction $18/13 \approx 1.385$ of Eq. (4.42). We have also plotted on Figure 4.17 this spectrum compensated by k^{-4}. It is obvious that we are extremely far from a plateau. This seems to rule out, in the framework of these EDQNM calculations, Eyink and Thomson's [88] prediction concerning the appearance of such a backscatter slighly above the crossover $s = 3.45$. The calculation for $s = 3.6$ is shown in Figure 4.18. Now backscatter increases, with the exponent being close to 3.7, and remains without any k^4 component. We have

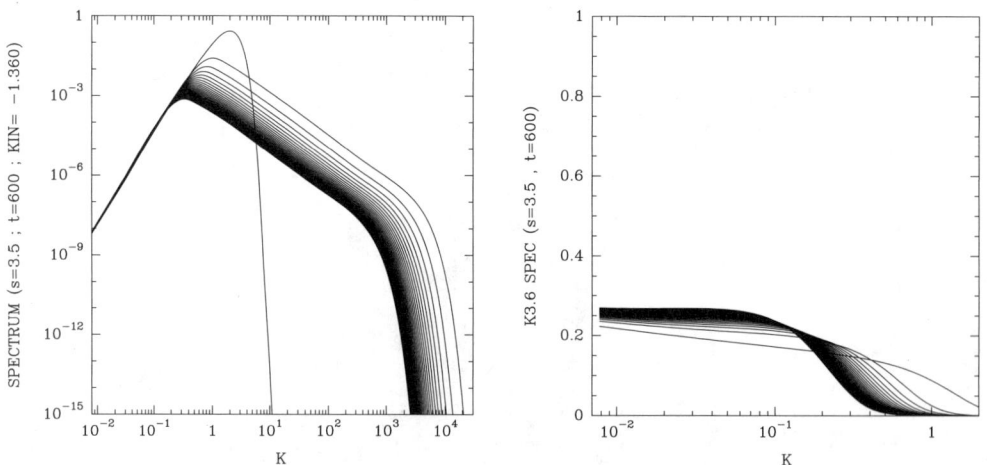

Figure 4.16. EDQNM decaying calculation for $s = 3.5$ showing time evolution of the kinetic-energy spectrum (left) and the spectrum multiplied by $k^{-3.6}$ (right).

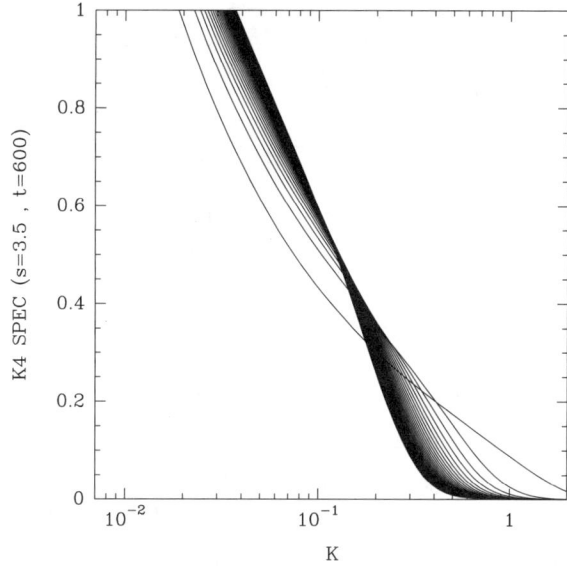

Figure 4.17. EDQNM decaying calculation for $s = 3.5$ showing time evolution of the kinetic-energy spectrum multiplied by k^{-4}.

$\alpha_E = 1.366$. For $s = 3.7$ (Figure 4.19), backscatter is more important, being close to $k^{3.8}$. We have $\alpha_E = 1.371$. For $s = 3.8$ (Figure 4.20), the backscatter is close to $k^{3.88}$, and we have $\alpha_E = 1.376$. We end with $s = 3.9$ (Figure 4.21), which is very close to the $s = 4$ case; now backscatter is $\sim k^{3.95}$, and $\alpha_E = 1.380$. The conclusion of this study is that, up to $s = 3$, permanence of large eddies holds, and the kinetic-energy decay is well described by Eq. (4.42). Backscatter appears gradually between $s = 3$ and $s = 4$. It does not bring a k^4 component to the spectrum; rather it yields a $k^{s'}$ spectrum, with s' slightly superior to s. Such a backscatter is in fact quite low up to $s = 3.4$ and is intensified above this value. In this respect, there is some crossover

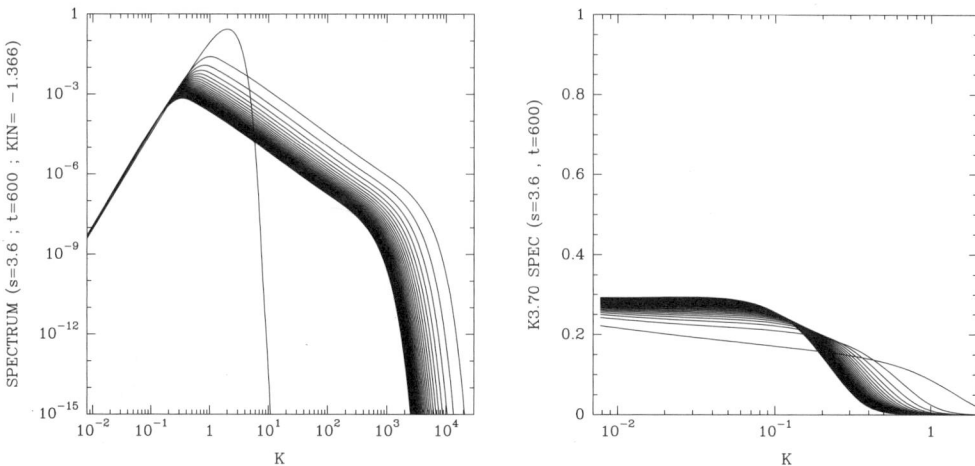

Figure 4.18. EDQNM decaying calculation for $s = 3.6$ showing time evolution of the kinetic-energy spectrum (left) and the spectrum multiplied by $k^{-3.7}$ (right).

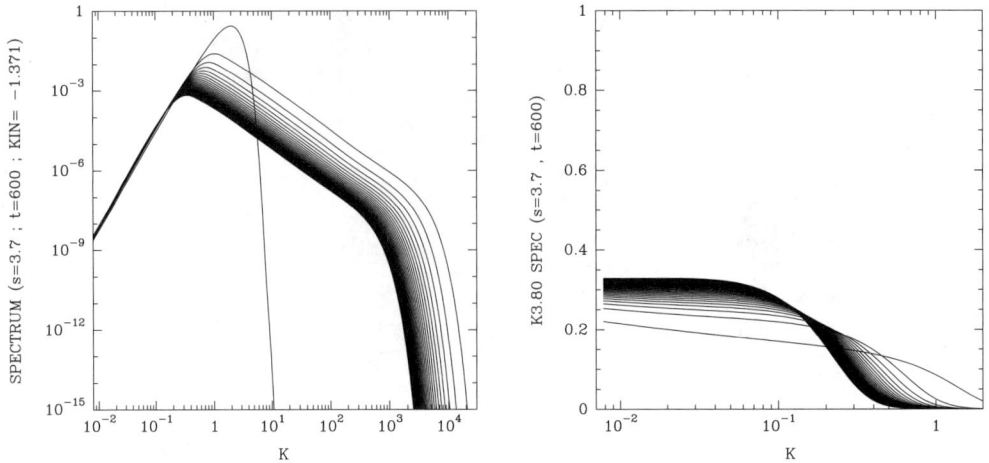

Figure 4.19. EDQNM decaying calculation for $s = 3.7$ showing time evolution of the kinetic-energy spectrum (left) and the spectrum multiplied by $k^{-3.8}$ (right).

between 3.4 and 3.5, but it is not as sharp as that proposed by Eyink and Thomson [88].

4.7.1 Finite-box size effects

Lesieur and Ossia [174] have also considered the case where the spectral peak is initially close enough to δk, so that confinement effects may further constrain the evolution of turbulence. In an experiment in liquid helium where turbulence decayed behind a grid, Stalp et al. [274] showed two stages in the kinetic-energy evolution: first a $t^{-1.2}$ law, followed by a t^{-2} law. They

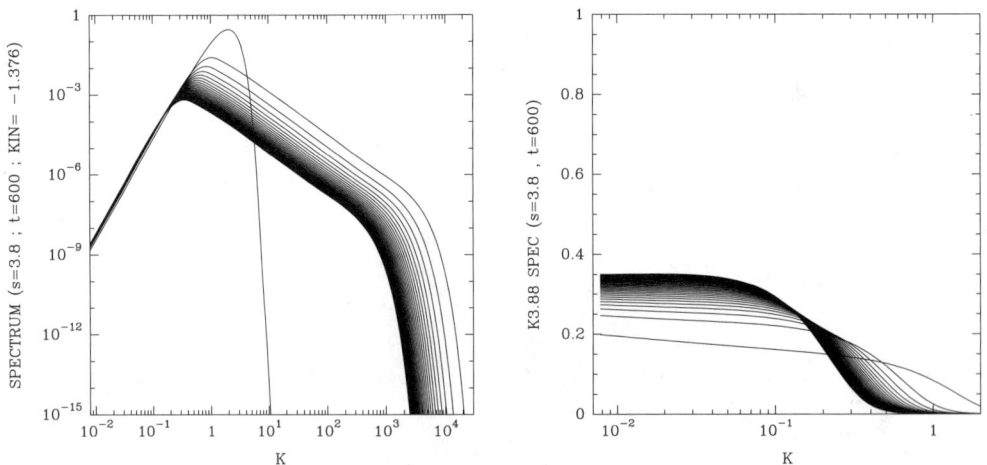

Figure 4.20. EDQNM decaying calculation for $s = 3.8$ showing time evolution of the kinetic-energy spectrum (left) and the spectrum multiplied by $k^{-3.88}$ (right).

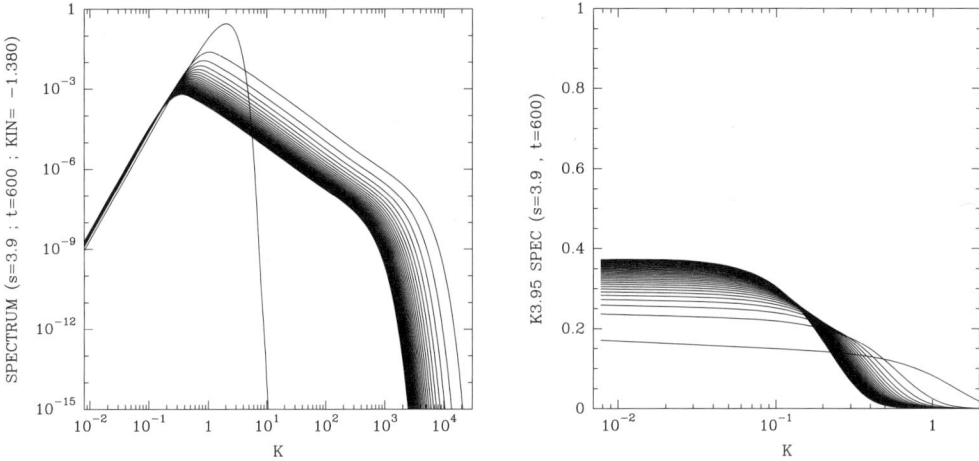

Figure 4.21. EDQNM decaying calculation for $s = 3.9$ showing time evolution of the kinetic-energy spectrum (left) and the spectrum multiplied by $k^{-3.95}$ (right).

interpreted the latter as corresponding to the decay of turbulence with a fixed integral scale. Indeed, let us assume the kinetic-energy spectrum is zero up to a constant k_i and decreases above k_i as $\epsilon^{2/3} k^{-5/3}$; then the kinetic energy is proportional to $\epsilon^{2/3}$. This yields $\alpha_E = 2$. In the EDQNM calculation, Lesieur and Ossia [174] take $s = 2$, $\delta_k = 0.125$, and $k_i(0) = 2$. Soon $E(k, t)$ forms a time-decaying spectrum close to $k^{-5/3}$ extending above δk (Figure 4.22 left). The time evolution of the exponent α_E is shown in Figure 4.22 (right). The exponent increases continuously from an initial value close to 1.2, and it is only at $t \approx 13,000$ that α_E seems to relax toward 2. At this time, the kinetic-energy spectrum is still close to $k^{-5/3}$ with, however, a rise at k_i. The corresponding Reynolds number (based on k_i) is of the order of a few thousand. Developments of this study have been carried out by Touil et al. [280].

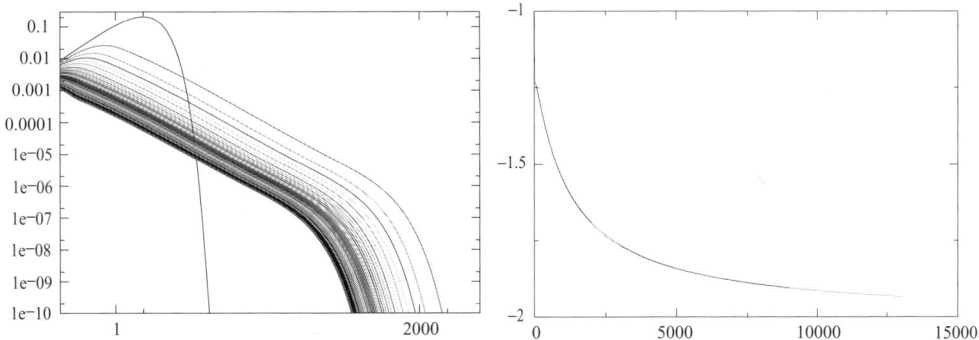

Figure 4.22. Time evolution in decaying EDQNM bounded turbulence of the kinetic-energy spectrum up to $t = 1,000$ (left) and of $-\alpha_E$ up to $t = 13,000$ (right).

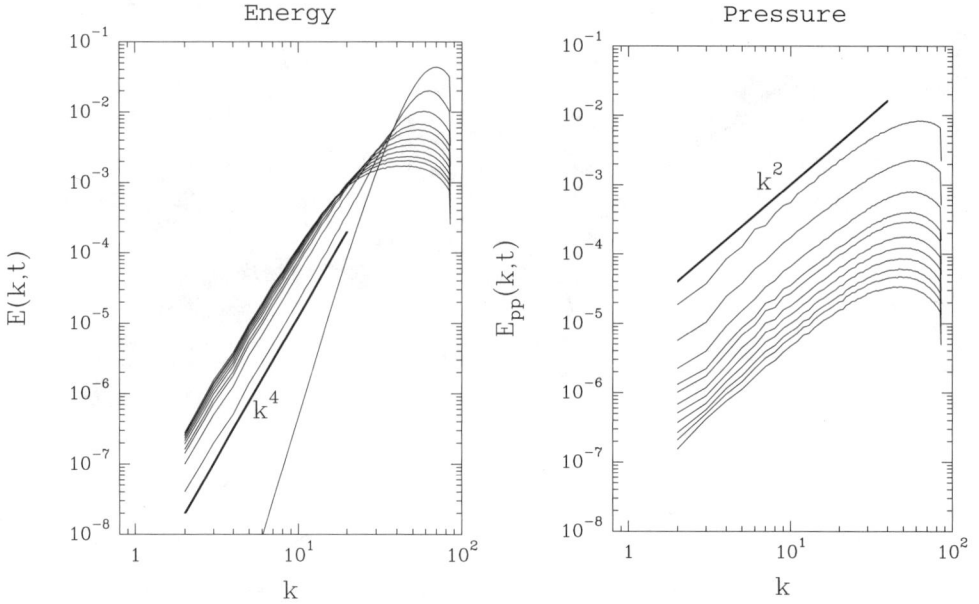

Figure 4.23. infrared kinetic-energy (left) and pressure (right) spectra in a 256^3 point simulation of isotropic decaying turbulence with $E(k, 0) \propto k^8$ for $k \to 0$. Time (from 0 to 20 initial turnover times) goes from top to bottom curve for the pressure. (From Lesieur et al. [171].)

4.8 Recent LES studies of decaying isotropic turbulence

The spectral-dynamic model was used by Lesieur et al. [171] and Ossia and Lesieur [226] to study the decay of isotropic three-dimensional turbulence. They looked at kinetic-energy and pressure infrared dynamics. Here, the kinetic-energy spectrum allowing us to determine the exponent m arising in the model is calculated by a three-dimensional spatial average in the computational box. The study of pressure spectra is particularly interesting for acoustic studies because they characterize the noise emitted.

In the results shown in the following, to have the widest infrared range available, the initial energy spectral peak is close to the cutoff. The number of collocation points in Fourier space will vary. The nonlinear terms are dealiased by the so-called $\frac{2}{3}$-rule. In the 256^3 point simulation, for instance, only 85 active Fourier modes are considered in each direction of Fourier space, between $k_{\min} = 1$ and $k_{\max} = 85$. The 256^3 point simulation presented on Figure 4.23 concerns an initial k^8 infrared kinetic-energy spectrum with a Gaussian velocity field. The calculation is run here up to 20 initial large-eddy turnover times. One sees that the kinetic-energy spectrum immediately picks up a k^4 behavior with a positive transfer, which confirms the existence of the k^4 backscatter. In fact, when the infrared kinetic-energy spectrum is proportional to k^4, the multiplying coefficient is equal to $I(t)/24\pi^2$, where $I(t)$, the

Loitzianskii integral, is defined by

$$I(t) = \int r^2 U_{ii}(r, t)d\vec{r}, \qquad (4.43)$$

and $U_{ii}(r, t)$ is the trace of the second-order velocity correlation tensor. This proves that the latter quantity is not strictly time invariant, although it does not vary much in the late stages of the computation. We remark that time invariance of Loitzianskii's integral, related by Landau and Lifchitz [155] to angular-momentum conservation of the flow,[10] yields a $t^{-10/7}$ time-decay law of the kinetic energy [corresponding to $s = 4$ in Eq. (4.42)], as was shown first by Kolmogorov [146]. Such a law may be recovered with a two-slope (4 and $-5/3$) model for the kinetic-energy spectrum (Comte-Bellot and Corrsin [50]), or with the assumption of a Karman–Howarth-type self-similar energy spectrum (Lesieur [170], p. 248), provided backscatter is neglected.

The infrared pressure spectrum of Figure 4.23, in contrast, follows immediately a k^2 law and decays rapidly (no pressure backscatter). Notice that such a law may be derived analytically using nonlocal expansions of the quasi-normal (or EDQNM[11]) pressure-spectrum equation. The theory predicts for the pressure spectrum

$$E_{pp}(k) = \left[\frac{8}{15}\int_0^{+\infty}\frac{E^2(q)}{q^2}dq\right]k^2, \qquad (4.44)$$

whereas in LES

$$E_{pp}(k, t) \approx \left[0.3\int_0^{k_C}\frac{E^2(q, t)}{q^2}dq\right]k^2. \qquad (4.45)$$

The latter law, which has been validated with compensated spectra, persists for a very long time (several hundred initial turnover times; see [226]) and is in fact independent of the infrared kinetic-energy spectrum behavior. Indeed LESs starting with an infrared k^2 kinetic-energy spectrum show the absence of energy backscatter in this case (see Figure 4.24) with a time-independent coefficient. The corresponding kinetic-energy decay law is Saffman's [247] $t^{-6/5}$ law. Meanwhile, the pressure still behaves according to the law (4.45). As was brought to our attention by Hill [124], a k^2 shape for the infrared pressure spectrum is implied without any approximation in Batchelor [19]. However, this result assumes that fourth-order correlations of velocity derivatives decrease to zero with sufficient rapidity as the separation goes to infinity. Such an assumption poses problems when one considers infrared energy spectra of

---

[10] In fact, such a principle does not hold exactly because of viscous-dissipation and boundary-conditions effects.

[11] Both theories are equivalent for the pressure, as was shown by Larchevêque [156].

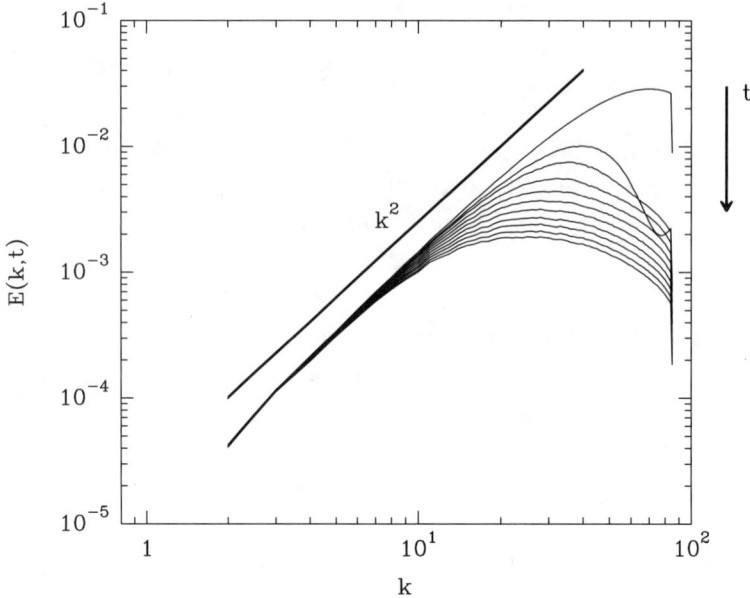

Figure 4.24. 256^3 point LES of decay of the kinetic-energy spectrum with $E(k, 0) \propto k^2$ for $k \to 0$. Time arrow goes from top to bottom. (From Lesieur et al. [171].)

various shapes. No such limitation exists for the closure result. Another great advantage of the closure here is that it provides an analytic expression for the pressure coefficient. Ossia and Lesieur [174] have determined in these 256^3 point LESs the time-decay laws of kinetic energy and pressure variance at very large times. The decay exponents are determined by a least-squares method over a time interval [30, 200]. They depend on the initial infrared exponent s of the kinetic-energy spectrum. Ossia and Lesieur found for $s = 4$ that $\alpha_E = 1.40$ (a value intermediate between the Kolmogorov prediction $10/7 = 1.43$ and Lesieur and Schertzer's 1.38 EDQNM prediction) and $\langle p'^2 \rangle = t^{-\alpha_P}$ with $\alpha_P = 2.90$; for $s = 2$ they obtain $\alpha_E = 1.22$ and $\alpha_P = 2.60$.

Notice that the values $\alpha_E = 1.40$ and $\alpha_P = 2.89$ have been found by Ossia [227] in a 512^3 point computation with $s = 4$. Notice also that the pressure variance time-decay law is not too far from that of the squared kinetic energy ($\alpha_P \approx 2\alpha_E$). Such a law had been found by Batchelor [19] with the quasi-normal approximation. The exponents found in the two cases for the kinetic energy do not saturate exactly at large times, as in the LES of Chasnov [38] using the plateau-peak model. Chasnov concludes that the exponents should eventually asymptote on the respective values 1.43 and 1.20, which would cancel the backscatter in the first case. However, these simulations at very large times might be affected by finite-box effects. These would prevent the integral scale of turbulence from growing and hence inhibit the spectral backscatter. The pressure spectrum, however, does not seem to be affected

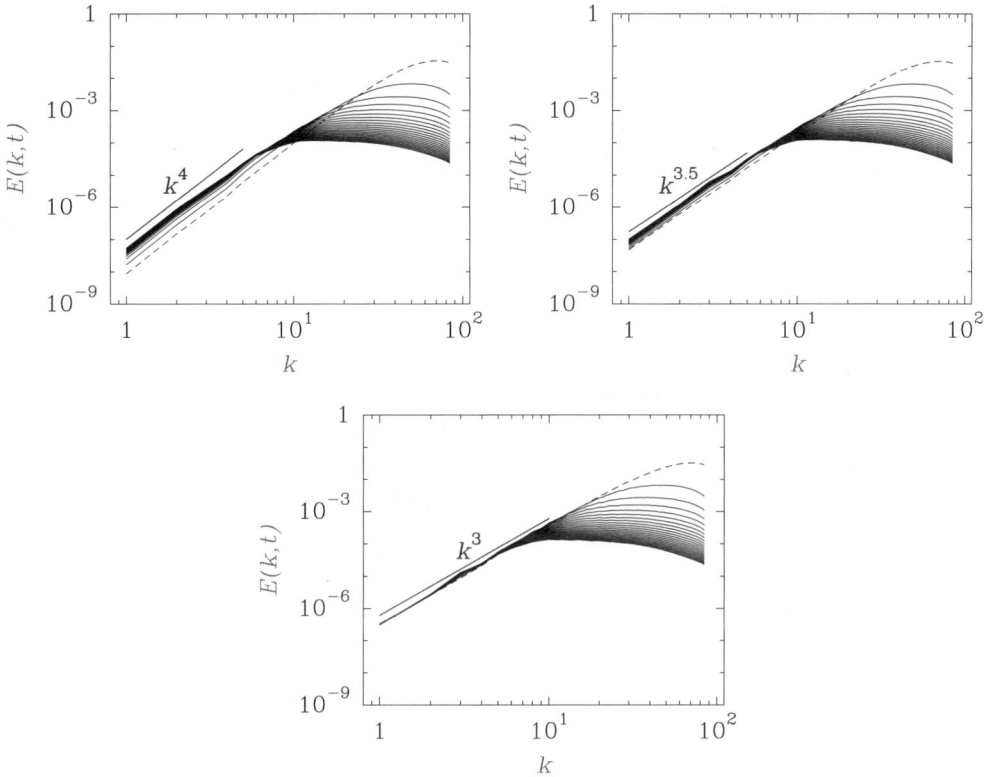

Figure 4.25. 256^3 LES of infrared kinetic-energy spectra in three-dimensional isotropic decaying turbulence with $E(k, 0)$ respectively as $\propto k^4, k^{3.5}, k^3$ for $k \to 0$. The spectra are shown from $t_0 = 0$ (dashed line) to $t_f = 200$ by $\Delta t = 8$. (From Ossia and Lesieur [226].)

by these problems. It is remarkable that, in the infrared pressure problem, closures of the quasi-normal/EDQNM type not only give good integral scaling on the kinetic-energy spectrum but also a correct order of magnitude for the coefficient.[12]

To finish with the Ossia and Lesieur [226] spectral-dynamic LES, let us mention the simulations with 256^3 collocation points and initial infrared exponents respectively equal to $s = 3$, $s = 3.5$, and $s = 4$ (see Figure 4.25). There is a quasi-permanence of big eddies for the case $s = 3$, as in the EDQNM prediction, but the decay exponent obtained is different: $\alpha_E = 1.36$ in the LES instead of 1.323 in the EDQNM. For $s = 3.5$, there is obviously no k^4 backscatter, and thus Eyink and Thomson's [88] prediction is again not satisfied. The LES yields here $\alpha_E \approx 1.38$, which is very close to the $18/13 \approx 1.385$ self-similar prediction given by Eq. (4.42) for $s = 3.5$ (recall that the EDQNM in this case yields $\alpha_E = 1.36$).

---

[12] Indeed, $8/15 \approx 0.53$, which should be compared with 0.3.

4.8.1 Ultraviolet pressure

A question of interest concerns the pressure behavior in the Kolmogorov $k^{-5/3}$ cascade. The quasi-normal analysis yields the Oboukhov–Batchelor spectrum (see in particular Batchelor [19])

$$E_{pp}(k) \sim \epsilon^{4/3} k^{-7/3}, \qquad (4.46)$$

which is an expression that can also be obtained dimensionally by assuming that the pressure spectrum $E_{pp}(k)$ is a function of ϵ and k only. Such a prediction is, however, controversial. Indeed, the LES of isotropic turbulence of Métais and Lesieur [205] indicates such a law for only a tiny range with, however, the right numerical value predicted by Monin and Yaglom [212]. In the LES of a temporal mixing layer using the spectral-dynamic model, Silvestrini et al. [268] found for the three-dimensional kinetic-energy spectrum an approximate average inertial exponent of $-5/3$ for $10 < k\delta < 40$, where $\delta(t)$ is the vorticity thickness. The pressure spectrum in this range displays a quite good $k^{-5/3}$ law.

5 Spectral LES for inhomogeneous turbulence

We present in this chapter applications of LES using a spectral eddy viscosity to two incompressible shear flows: a temporal mixing layer and a plane channel.

5.1 Temporal mixing layer

5.1.1 Plateau-peak model

We apply the (nondynamic) plateau-peak model to an incompressible temporal mixing layer. The flow is periodic in the streamwise and spanwise directions, and free-slip conditions are assumed on the lateral boundaries. The numerical code is pseudospectral in the three dimensions of space. There is no molecular viscosity. Turbulence is initiated by a hyperbolic-tangent velocity profile[1]

$$\bar{u}(y) = U \ \tanh \frac{y}{\delta_0}, \tag{5.1}$$

where $\delta_i = 2\delta_0$ is the initial vorticity thickness. The length of the domain corresponds to $4\lambda_a$, where

$$\lambda_a = 14\delta_0 = 7\delta_i \tag{5.2}$$

is Michalke's [208] inviscid, most-unstable wavelength.

Three-dimensional forcing
A small, three-dimensional, white-noise, random perturbation close to isotropy is first superposed to this basic profile. Figure 5.1 (top) shows a perspective

---

[1] This is a good approximation in the laminar case or for turbulent mean velocity.

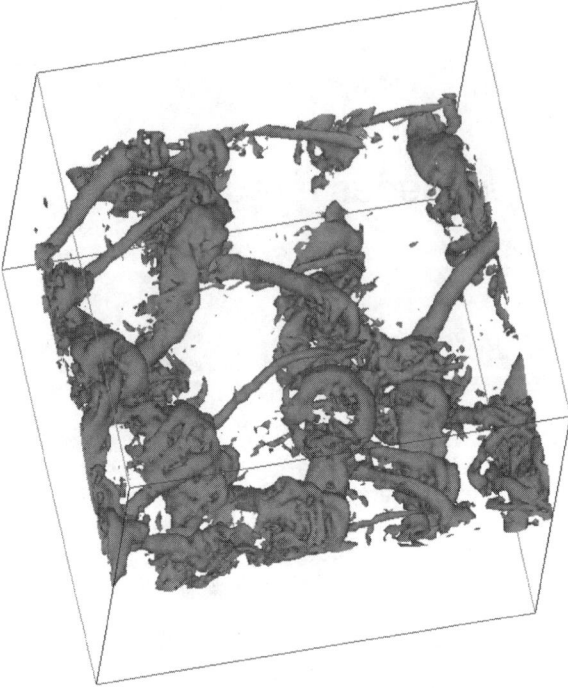

Figure 5.1. Vorticity modulus at $t = 14\delta_i/U$ in the LES of a temporal mixing layer forced initially. (Top) three dimensionally. (Bottom) quasi-two-dimensionally. (Courtesy G. Silvestrini.)

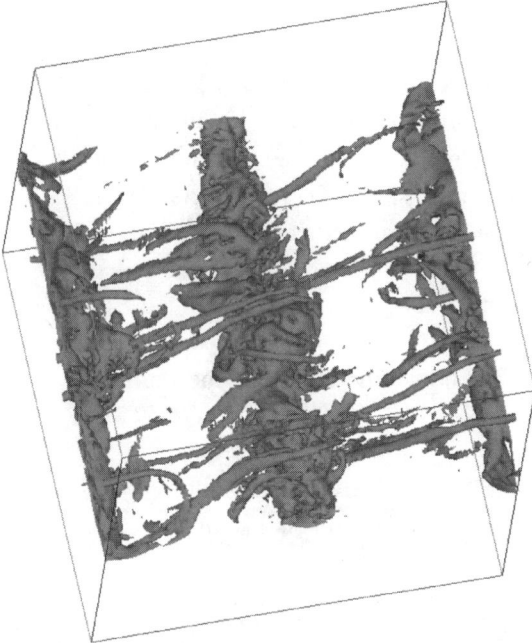

view of the vorticity modulus at $t = 14\delta_i/U$. It displays evidence for helical pairing, where vortex filaments oscillate out of phase in the spanwise direction and reconnect, yielding a vortex-lattice structure. This was previously found in the DNS of Comte et al. [51, 52] with the same initial conditions.

Quasi-two-dimensional forcing

If the perturbation is quasi-two-dimensional, the mixing layer evolves into a set of large quasi-two-dimensional Kelvin–Helmholtz (KH) vortices, both of which undergo pairing and stretch intense longitudinal hairpin vortices in the stagnation regions between them. Figure 5.1 (bottom) displays the vorticity-modulus field at $t = 14\delta_i/U$. The circulation of longitudinal hairpin vortices is of the order of that of the basic KH vortices These longitudinal vortices have long been observed experimentally in spatially growing incompressible mixing layers (see, e.g., Bernal and Roshko [23]). They are due to the stretching of vorticity within the stagnation regions between the large rollers, following mechanisms reviewed in Lesieur ([170]; see also Corcos and Lin [55] and Neu [218]). We consider the Euler vorticity equation

$$\frac{D\vec{\omega}}{Dt} = \vec{\nabla}\vec{u} \otimes \vec{\omega} = \bar{\bar{S}} \otimes \vec{\omega} + \frac{1}{2}\vec{\omega} \times \vec{\omega} = \bar{\bar{S}} \otimes \vec{\omega}. \qquad (5.3)$$

The deformation tensor $\bar{\bar{S}}$ is real and symmetric. It has therefore real eigenvalues, and the eigenvectors (principal axes of deformation) form an orthonormal frame. We suppose that the vorticity in the stagnation region is weak initially and that the eigenvalues of $\bar{\bar{S}}$, ranked as $s_1 \geq s_3 \geq s_2$, will not change during the evolution. Given that $s_1 + s_2 + s_3 = 0$ because of incompressibility, we get $s_1 > 0$, $s_2 < 0$. We assume also that the eigenvectors are time independent. Working in the orthonormal frame they form, we have

$$\frac{D\omega_1}{Dt} = s_1\omega_1, \quad \frac{D\omega_2}{Dt} = s_2\omega_2, \quad \frac{D\omega_3}{Dt} = s_3\omega_3.$$

Vorticity will be stretched in the direction of the first principal axis and compressed in the direction of the second. In fact the flow in this stagnation region is not far from a plane pure deformation, and thus $s_3 \approx 0$ (spanwise direction) and the first axis of deformation is offset by $45°$ with respect to the direction of the mean flow. This justifies the assumption of time independence for the eigenvectors.

It is remarkable that the present LESs, done at a quite low resolution (using pseudo-spectral methods and 96^3 Fourier modes), are able to capture the longitudinal vortices, which are at quite small scales. The DNS studies of Comte et al. [52] at a comparable resolution were unable to find organized, intense longitudinal vortices in the quasi-two-dimensional forcing case because their molecular Reynolds number ($U\delta_i/\nu = 100$ initially) was too low. It is only at a much higher resolution that DNS can capture these vortices (Rogers and Moser [245]). This is an example where LESs provide an excellent tool for capturing not only large but also small-scale vortices. We will return to other aspects of temporal or spatial mixing layers throughout the rest of the book.

5.1.2 Spectral-dynamic model

The same simulations as just described have been carried out by Silvestrini et al. [268] using the spectral-dynamic model. The spectral exponent m of the model is here computed with the aid of the three-dimensional kinetic-energy spectrum calculated through a volume average in the domain, and m is determined using a least-squares method applied in the range $[k_C/2, k_C]$.

Three-dimensional forcing

For the three-dimensional initial forcing, m decreases from a value of 9 to about 2 at $t \approx 20\delta_i/U$, the value at which it saturates. Helical pairing is recovered at $t = 14\delta_i/U$ as for the plateau-peak-based LES already presented. At the end of the simulation ($t = 60\delta_i/U$), the mixing layer is highly three-dimensional. Statistics for mean and rms velocity profiles as well as Reynolds stresses normalized by U and the vorticity thickness $\delta(t)$ are in very good agreement with experiments on unforced, spatially growing mixing layers by Bell and Mehta [21]. In fact, Bell and Mehta [21] obtained a self-similar regime at a distance of about $250\delta_i$ from the splitter plate. Because the vortices in the spatial case travel at a velocity $(U_1 + U_2)/2$, events labeled by some time $N\delta_i/U$ in the temporal problem will in the spatial case correspond to a distance $N[(U_1 + U_2)/(U_1 - U_2)]\delta_i$. Bell and Mehta [21] have a velocity ratio $(U_1 - U_2)/(U_1 + U_2)$ of 0.25. Then the time of $60\delta_i/U$ in the temporal problem corresponds to a downstream distance of $240\delta_i$ in Bell and Mehta's [21] experiment, which is not far from the value of 250 they propose.

The kinetic-energy spectra (normalized by U^2 and $k\delta$) of the temporal mixing layer collapse well in the small scales between $t = 50\delta_i/U$ and $t = 60\delta_i/U$ with an acceptable Kolmogorov $k^{-5/3}$ spectrum on about half a decade above $k\delta = 10$. This differs from the experiment, where the $k^{-5/3}$ in which spectra cover a wider range and should be attributed to a defect of the spectral-dynamic model already noticed by Ossia and Lesieur [226] for isotropic turbulence.

As stressed in Chapter 4, the pressure spectrum of the temporal mixing layer follows a $k^{-5/3}$ law in the same range as the energy spectrum.

Quasi-two-dimensional forcing

For the temporal mixing layer with a quasi-two-dimensional initial forcing, m has the same behavior as in the three-dimensional forcing case.

One observes rollup of quasi-two-dimensional KH vortices at $t = 10\delta_i/U$. The first pairing starts at $t = 14\delta_i/U$ with the merging of central vortices in the computational box and is finished with the merging of the two exterior vortices at $t = 30\delta_i/U$. Intense, thin longitudinal vortices are also stretched longitudinally. At $t = 45\delta_i/U$ the central KH vortex begins to oscillate in

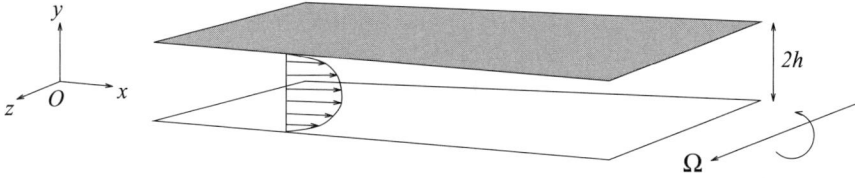

Figure 5.2. Schematic view of a plane channel.

the spanwise direction, which triggers the second pairing. At the end of the simulation ($t = 85\delta_i/U$), there is one large spanwise KH vortex stretching another large longitudinal vortex. However, compared with the statistical data of Bell and Mehta [21], it is clear that the simulation has yet to reach self-similarity.

5.2 Incompressible plane channel

We now show how the spectral-dynamic model may be applied to an incompressible turbulent Poiseuille flow between two infinite parallel flat plates. A schematic view of the channel is presented in Figure 5.2. A rotation axis oriented in the spanwise direction is indicated for further applications, but rotation is inactive right now. The channel has a width $2h$, and we define the macroscopic Reynolds number by $Re = 2hU_{\mathrm{m}}/\nu$, where U_{m} is the bulk velocity (integral of the mean velocity across the channel divided by $2h$). We assume periodicity in the streamwise and spanwise directions.

5.2.1 Wall units

Let us recall the so-called wall units, for these are very useful when turbulence has developed. The friction velocity v_*, defined by setting the mean stress at the wall equal to ρv_*^2, satisfies

$$v_*^2 = \nu \left[\frac{d\langle u \rangle}{dy} \right]_{y=0}. \tag{5.4}$$

The velocities will be normalized by v_* and denoted $u_i^+ = u_i/v_*$. We also define a viscous thickness

$$l_{\mathrm{v}} = \frac{\nu}{v_*}, \tag{5.5}$$

which is characteristic of motions very close to the wall that are dominated by viscosity,[2] so that the spatial scales will be normalized by l_{v} and denoted $x_i^+ = x_i/l_{\mathrm{v}}$. Note that $h^+ = v_* h/\nu$ defines a microscopic Reynolds number

---

[2] At a Reynolds number high enough, this scale is comparable to the Kolmogorov dissipative scale previously introduced.

based on the friction velocity. Let y be the distance perpendicular to the wall. We have seen above that substituting Taylor-series expansions of the velocity components in powers of y^+ close to the wall leads to the well-known result that u^+ and w^+ scale like y^+, whereas v^+ scales like $y^{+}.^2$ The mean longitudinal velocity profile also scales like y^+ – a behavior that persists up to about $y^+ = 4$–5, which characterizes the width of the viscous region. DNS and LES show that this region is certainly not laminar but is strongly marked by a system of high- and low-longitudinal velocity streaks, which we are going to discuss.

5.2.2 Streaks and hairpins

Since the experimental observations of Kline et al. [144] of a turbulent boundary layer without a pressure gradient above a flat plate, it has been well established that coherent structures in the form of streaks of high- and low-longitudinal velocity fluctuations exist up to about 50 wall units from the wall. Their average length is of the order of 500 wall units, but low-speed streaks are longer than high-speed ones. In fact, velocity streaks had been observed by Klebanoff et al. [143] in a celebrated paper related to transition in a boundary layer forced upstream by a vibrating ribbon. Klebanoff associated the streaks (which he could detect with probes) with a system of longitudinal hairpins traveling downstream and pumping between their legs lower fluid slowed by the wall. This model explains the formation of low-speed streaks in the "peaks" of the hairpins and high-speed streaks in the "valleys." The system of streaks in a turbulent channel was recovered numerically in a LES of Moin and Kim [210] using Smagorinsky's model with wall laws.

5.2.3 Spectral DNS and LES

We will present turbulent channel DNS and LES taken from the work of Lamballais [151] and Lamballais et al. [153]. Calculations are carried out at constant U_m, which replaces the forcing provided by the mean longitudinal pressure gradient. They are initiated by a parabolic laminar profile perturbed by a small, three-dimensional random noise and are pursued up to complete statistical stationarity. The numerical code used combines pseudo-spectral methods in the streamwise and spanwise directions and compact finite-difference schemes of sixth order in the transverse direction. The subgrid model is the spectral-dynamic eddy viscosity computed via two-dimensional kinetic-energy spectra calculated at each time step by spatial averages in planes parallel to the wall.[3] Therefore, the exponent m in the eddy viscosity depends on y and t. This spectral eddy viscosity is implemented spectrally in the directions parallel to the wall and in physical space in the transverse direction. This is a very precise code of accuracy comparable to a spectral method at equivalent resolution, as shown by the comparison of DNS at $h^+ = 162$ with spectral DNS of Kuroda [149] at $h^+ = 150$ (see Lesieur [170], p. 118). We reproduce this picture in Figure 5.3. We see in Figure 5.3(a) that the logarithmic range starts at $y^+ = 30$. Figure 5.3(b) presents the rms velocity profiles as a function of y^+. It confirms the strong u' production close to the wall with a peak at $y^+ = 12$, which is obviously the signature of the high- and low-speed streaks discussed before. Figure 5.3(e) (Reynolds stresses) and 5.3(d) (rms pressure fluctuations) have a higher peak ($y^+ \approx 30$). It might correspond to the tip of hairpin vortices traveling above the low-speed streaks, as proposed by Lesieur [170], whose evidence will be presented in animations later on in the book. Figure 5.3(f) corresponds to rms vorticity fluctuations and is interpreted in Lesieur ([170], p. 117), who writes (footnotes added), "[The figure] shows the r.m.s. vorticity fluctuations (a quantity very difficult to measure precisely experimentally[4]). It indicates that the maximum vorticity produced is spanwise and at the wall.[5] The vorticity perpendicular to the wall is about 40% higher than the longitudinal vorticity in the region $5 < y^+ < 30$, which shows only a weak longitudinal vorticity stretching by the ambient shear." We next present two LESs using the spectral-dynamic model at $Re = 6,666$ ($h^+ = 204$, case A) and $Re = 14,000$ ($h^+ = 389$, case B). They are, respectively, subcritical and supercritical with reference to the linear-stability analysis of the Poiseuille profile. In the two simulations the grid is refined close to the wall to simulate accurately the viscous sublayer. Figure 5.4 shows for case A the time-averaged exponent m arising in the energy spectrum at the cutoff as a function of the distance to the wall y^+. Regions where $m > 3$ correspond to a zero eddy viscosity and hence a direct numerical simulation. This is the case in particular close to the wall, up to $y^+ \approx 12$ where we know that longitudinal velocity fluctuations are very intense, owing to the low- and high-speed streaks. Therefore, and because the first point is very close to the wall ($y^+ = 1$), such LESs have the interesting property of becoming a DNS in the vicinity of the wall, enabling us to capture events that occur in this region. Figure 5.5 shows (in semilogarithmic coordinates) the mean velocity profile in case A compared

---

[3] It was determined by Lamballais [151] that the replacement of such a spectrum by a fictitious three-dimensional spectrum using isotropy relations (when spectra decrease as a power law) did not change the results significantly.

[4] This is rapidly changing with the impressive development of digital particle image velocimetry techniques.

[5] It corresponds in fact to a steepening of du/dy at the wall under the high-speed streaks resulting from a kind of squashing of the boundary layer on the wall as the fluid descends. It is in these regions that viscous friction is produced.

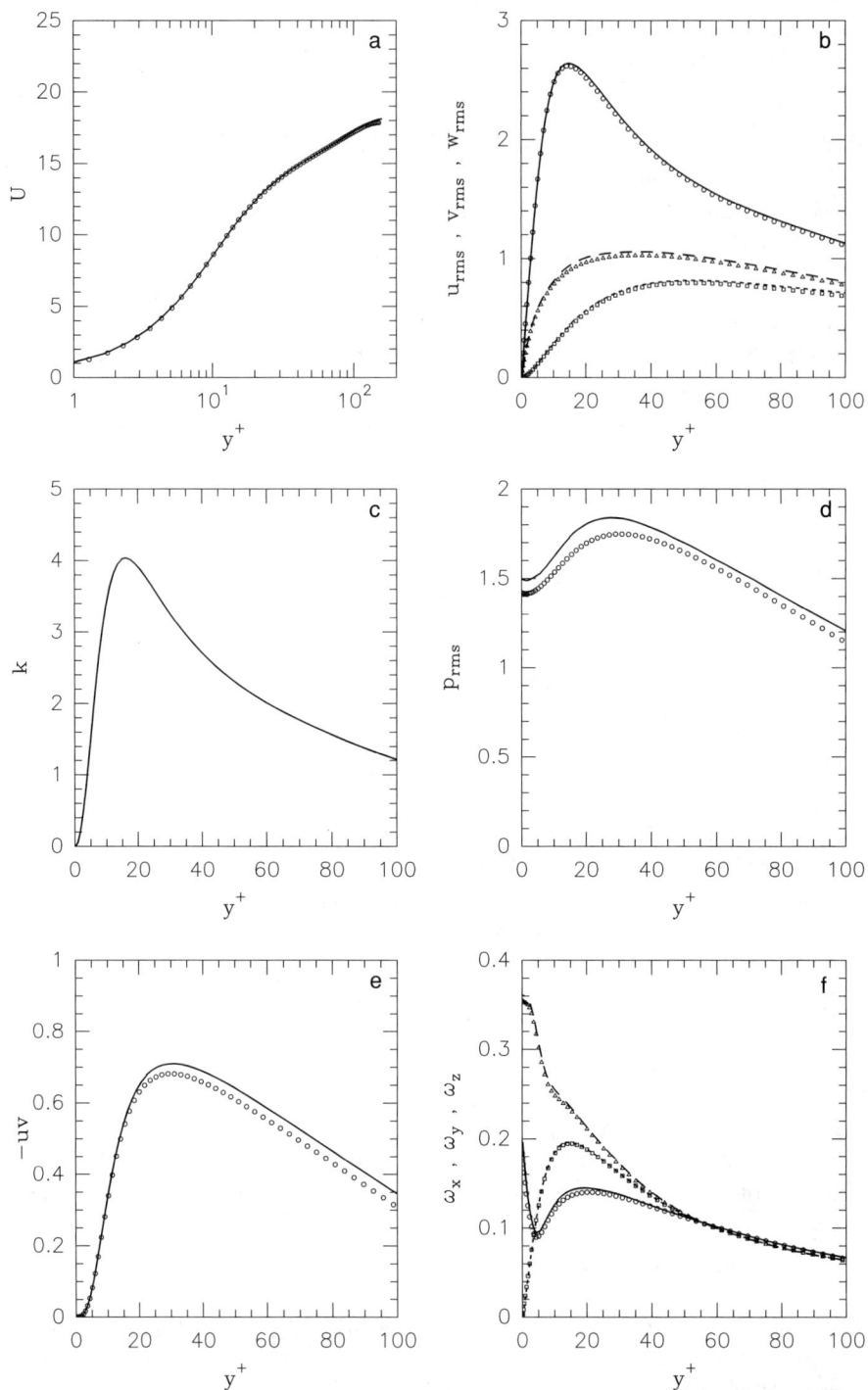

Figure 5.3. Statistical data obtained in DNS of a turbulent channel flow by Lamballais (straight line) and Kuroda (symbols): (a) mean velocity, (b) rms velocity fluctuations (respectively, from top to bottom, longitudinal, spanwise, and vertical), (c) kinetic energy, (d) rms pressure fluctuation, (e) Reynolds stresses, and (f) rms vorticity (from top to bottom, spanwise, vertical, and longitudinal). (From Lesieur [170]; courtesy Kluwer.)

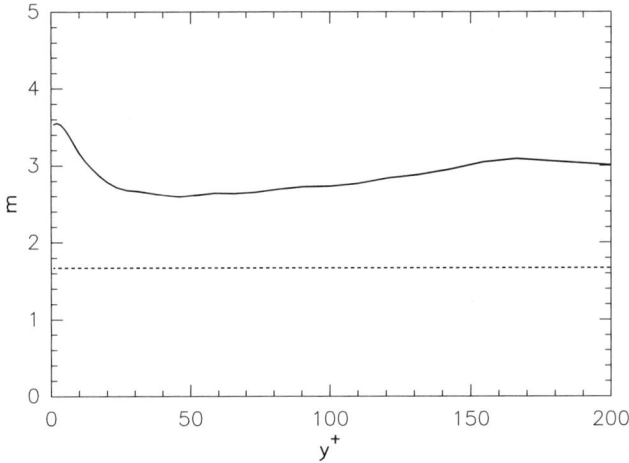

Figure 5.4. Spectral-dynamic LES of the channel flow (case A) showing time-averaged exponent $m(y^+)$ of the kinetic-energy spectrum at the cutoff.

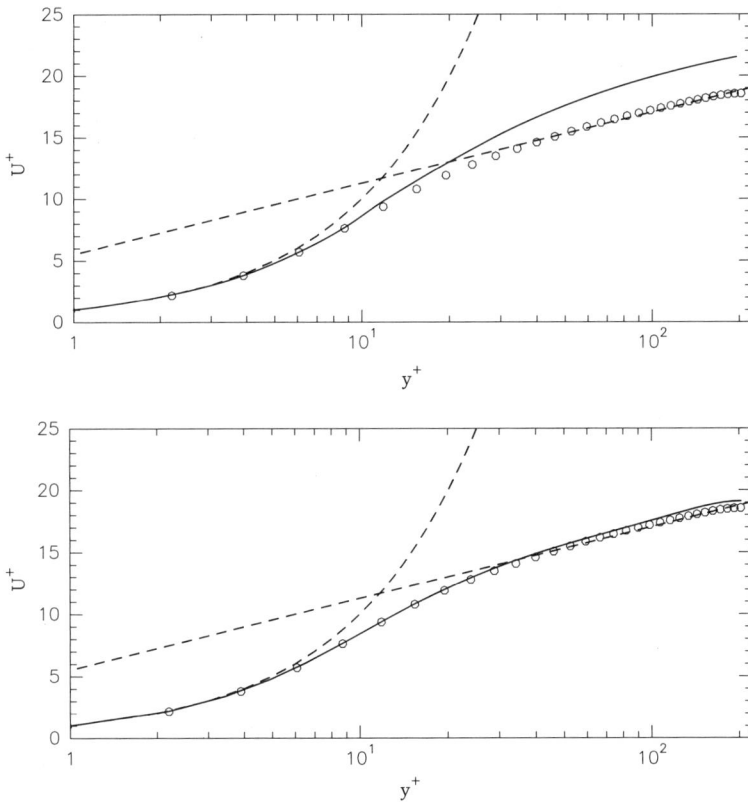

Figure 5.5. Mean velocity profiles in wall units. Lines, spectral eddy-viscosity-based simulations ($Re = 6,666$); symbols, Piomelli ([236], $Re = 6,500$). (Top) $m = 5/3$ ($h^+ = 181$); (bottom) spectral-dynamic model ($h^+ = 204$). The dashed straight line corresponds to the universal logarithmic mean velocity profile $\langle u \rangle = 2.5 \ln y^+ + 5.5$.

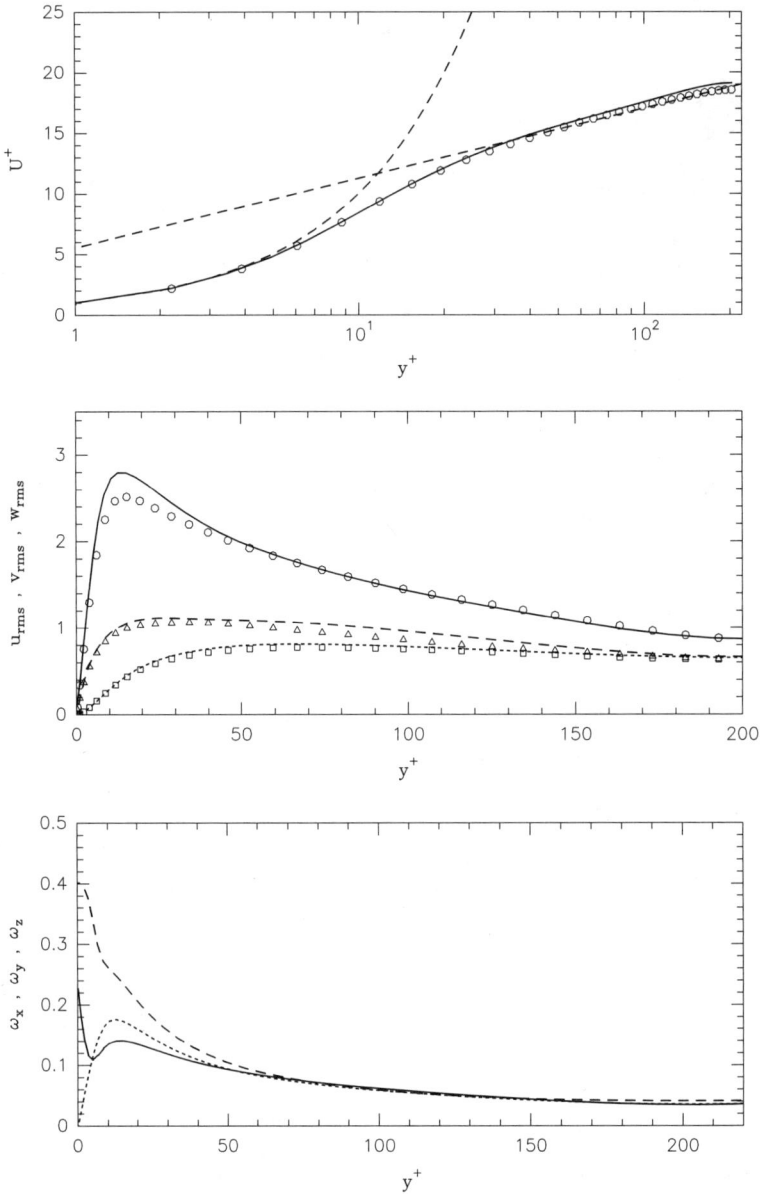

Figure 5.6. Spectral-dynamic LES. Same as Figure 5.5 with the rms velocity and vorticity fluctuations.

with the LES of Piomelli [236] using the dynamic model of Germano and co-workers [108, 109]. The latter is known to agree very well with experiments at these low Reynolds numbers. The simulation using the spectral-dynamic model (bottom part of the figure) coincides with Piomelli's, yielding a correct value of the additive constant 5.5 in the logarithmic velocity profile. However, the LES using the classical spectral-cusp model with $m = 5/3$ (top of figure) gives an error of 100% for this constant. The dashed parabola corresponds to the linear profile at the wall, which is exact up to 4 wall units. Figure 5.6 shows

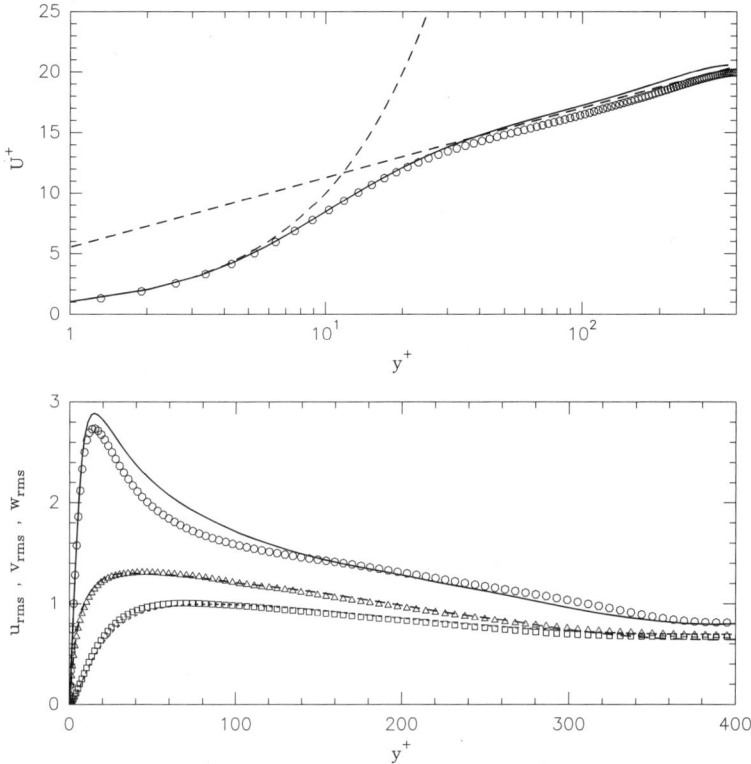

Figure 5.7. Turbulent channel flow comparisons of the spectral-dynamic model (solid lines, $h^+ = 389$) with the DNS of Antonia et al. ([7]; symbols, $h^+ = 395$). (Top) mean velocity; (bottom) rms velocity components.

for case A the mean velocity (same as in preceding figure) and rms velocity fluctuations compared with the dynamic-model predictions of Piomelli [236]. The agreement of rms velocities is still very good, with a correct prediction of the peak in longitudinal velocity fluctuations. For the supercritical case, the LESs of case B are in very good agreement with a DNS at $h^+ = 395$ carried out by Antonia et al. [7] both for the mean velocity and the rms velocity components. They are shown in Figure 5.7. Notice that in this case the LES allows us to reduce the computational cost by a factor of the order of 100. Notice also that the extent of the linear-velocity profile range close to the wall has slightly increased (from 4 to 5) with the Reynolds number. We present finally in Figure 5.8 a map of the vorticity modulus at the same threshold for cases A and B. The flow goes from left to right. It is clear that the LES does reproduce features expected from turbulence at higher Reynolds number and displays much more vortical activity in the small scales than the DNS.

5.2.4 Channel pdfs

We turn back to the DNS of the channel at $h^+ = 162$ (Figure 5.3), for which we present various probability density functions (pdfs) of pressure and velocity

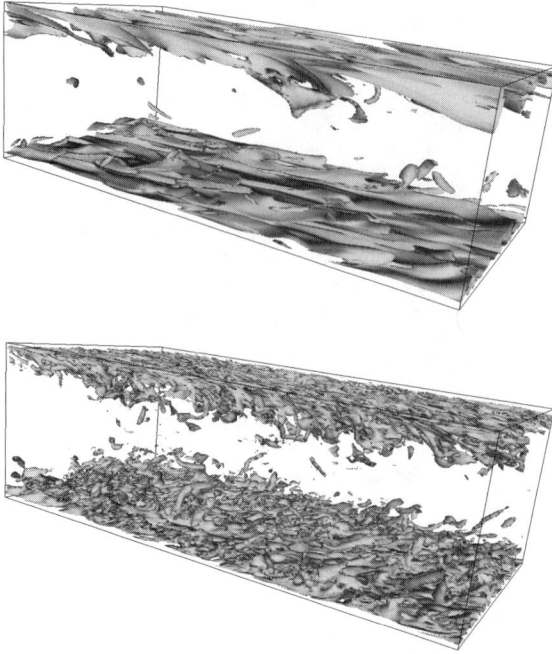

Figure 5.8. Turbulent plane channel vorticity modulus. (Top) DNS ($h^+ = 165$); (bottom) LES using the spectral-dynamic model ($h^+ = 389$). (From Lamballais [151].)

taken from Lamballais et al. [153]. They are obtained by temporal averaging as well as averaging in planes parallel to the wall. Consideration of both sides of the channel allows us to double the number of statistical samples. We have already seen that, in a turbulent flow, coherent vortices are generally characterized by a high vorticity modulus and a low pressure. Métais and Lesieur [205] showed that the pressure pdf was skewed in isotropic turbulence with a quasi-exponential tail in the lows and a Gaussian one in the highs. In contrast, the pdf of any vorticity component is symmetric with exponential-like tails. Analogous results were found in DNS of a mixing layer by Comte et al. [52] and, for the pressure, in experiments of turbulence between two counterrotating disks [91]. In all these cases, the coherent vortices are clearly identified, and one generally relates the skewed pressure pdf to the existence of vortices.[6] It is thus of interest to carry out the same study in a plane turbulent channel to see how the various pdfs of pressure and velocity react to the existence of streaks and ejected hairpins. All the pdfs presented here concern the fluctuations with respect to the mean, and the argument is normalized by the rms value. They are plotted in semilog coordinates, and the dashed parabola indicates a Gaussian distribution of variance 1.

Figure 5.9 shows the pdfs of pressure at the wall very close to it in the viscous region ($y^+ = 2.5$), where the streaks are maximum ($y^+ = 12$), and

---

[6] Indeed, the pressure signal will exhibit a strong undershoot within the vortex, inducing high probabilities for low-pressure values. However, as we have seen in Chapter 2, a Gaussian velocity field may contain large weak vortices, implying an asymmetric pressure pdf.

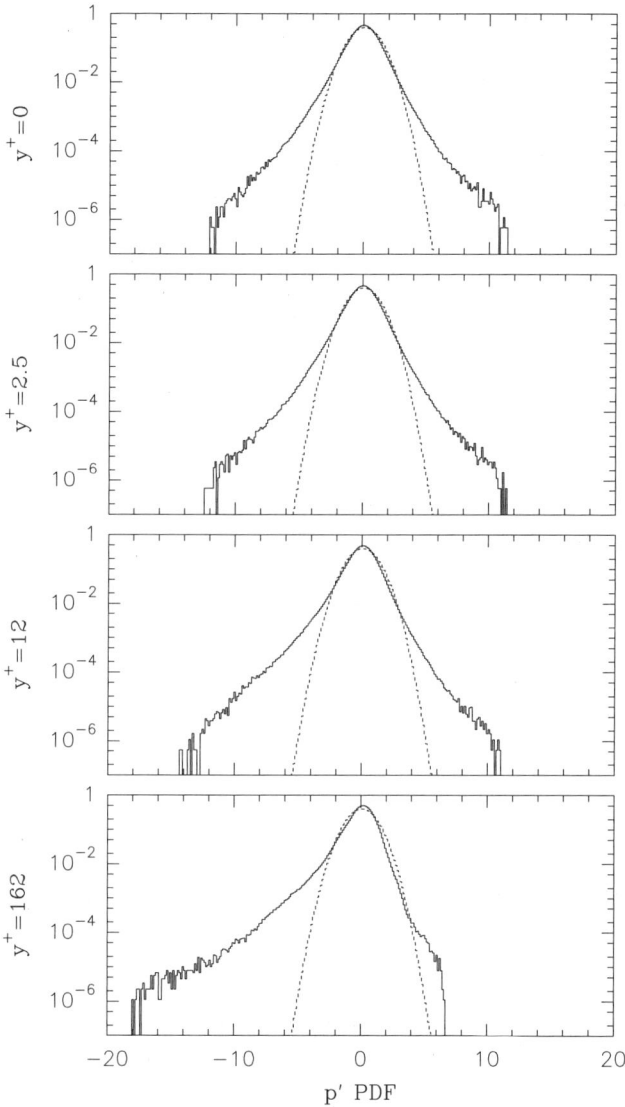

Figure 5.9. Channel pressure fluctuation pdfs at different distances from the wall. (Courtesy *Phys. Rev. E.* and E. Lamballais.)

in the core of the flow close to the channel center ($y^+ = 162$). Figures 5.10, 5.11, and 5.12 present, respectively, the longitudinal (u'), transverse (v'), and spanwise (w') velocity fluctuations at $y^+ = 2.5$, 12, and 162. The pressure pdf at the wall ($y^+ = 0$) is totally symmetric, with exponential wings. There is no trace of any kind of vortical organization. Such a pressure pdf had already been determined in analogous simulations by Kim et al. [142] but with a smaller number of statistical samples. At $y^+ = 2.5$, the pressure is very close to the distribution at the wall, and it is still difficult to see any trace of vortices. The longitudinal velocity is highly intermittent at high speeds

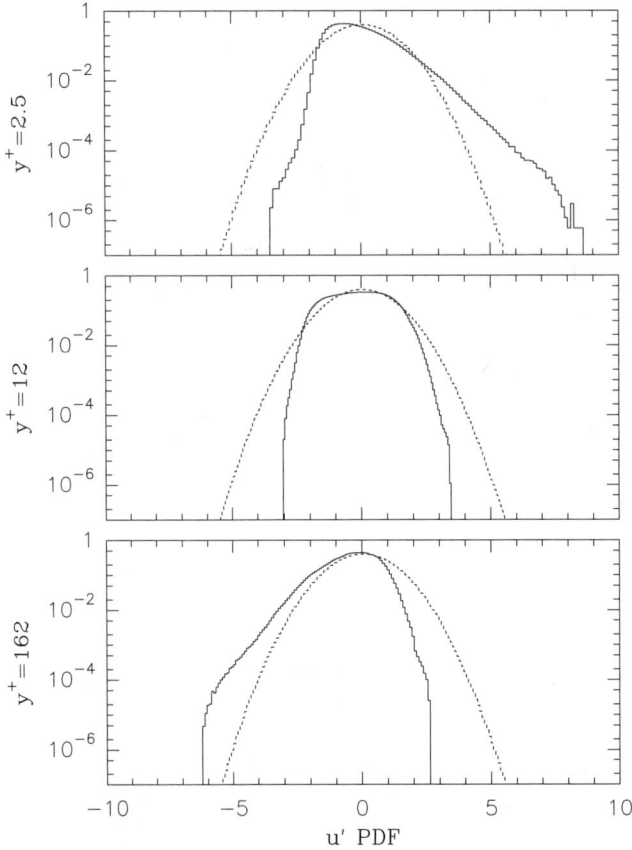

Figure 5.10. Channel longitudinal velocity pdf at different distances from the wall. (Courtesy *Phys. Rev. E.* and E. Lamballais.)

and "sub-Gaussian" at low speeds. Thus, high-speed streaks are much more intense and intermittent than the low-speed ones close to the wall. Such a distribution for u' at the wall is responsible for the positive skewness of u' measured in the experiments of Comte-Bellot [49] and recovered in the DNS of Kim et al. [142]. The skewness of u' is here defined as

$$\frac{\int_{-\infty}^{+\infty} u'^{3} P(u')du'}{\langle u'^{2}\rangle^{3/2}}, \tag{5.6}$$

where $P(u')$ is the pdf of u'. Another consequence of these abrupt excursions of positive longitudinal velocity in the high-speed streaks close to the wall is the creation of intense excursions of spanwise vorticity (and hence of drag) at the wall just underneath, which have the same sign as the basic vorticity. This point was noticed by Ducros et al. [81] in LES of a weakly compressible boundary layer spatially developing above a flat plate, which will commented on later in the book. Still, at $y^{+} = 2.5$, the pdf of v' is very intermittent with faster descents than ascents. This is in agreement with the fact that the

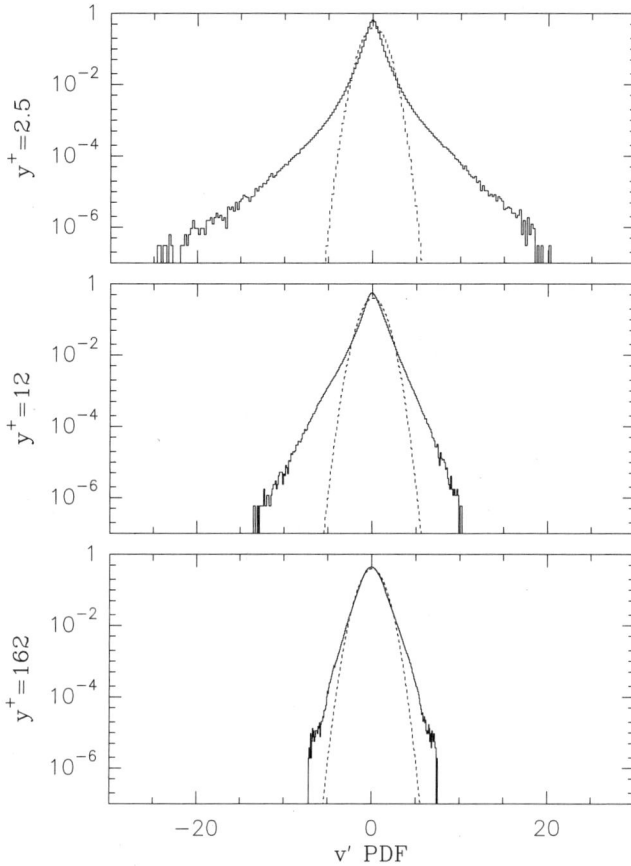

Figure 5.11. Channel vertical velocity pdf at different distances from the wall. (Courtesy *Phys. Rev. E.* and E. Lamballais.)

flow upwells in the low-speed regions and sinks in the high-speed regions. The spanwise velocity w' is extremely intermittent. It should be symmetric because of the mirror symmetry of turbulence with respect to planes parallel to (x, y). In fact, such a symmetry does not hold exactly owing to a lack of spanwise extent of the domain. In the core of the streaks ($y^+ = 12$), u' has no intermittency at all, since both sides of the pdf are sub-Gaussian. The vertical velocity v' is weakly asymmetric, with a preference still favoring descents over ascents. The spanwise velocity w' is not very intermittent. The pressure becomes asymmetric with more intermittency in the troughs than in the peaks, which are still exponential.

In the central region of the channel ($y^+ = 162$), the pressure pdf resembles skewed distributions encountered in isotropic turbulence or free-shear flows. Visualizations and animations of the vorticity and Q fields do show in fact that large, asymmetric, hairpin-shaped, quasi-longitudinal vortices are still carried by the flow, and their existence certainly explains the pressure distribution. The

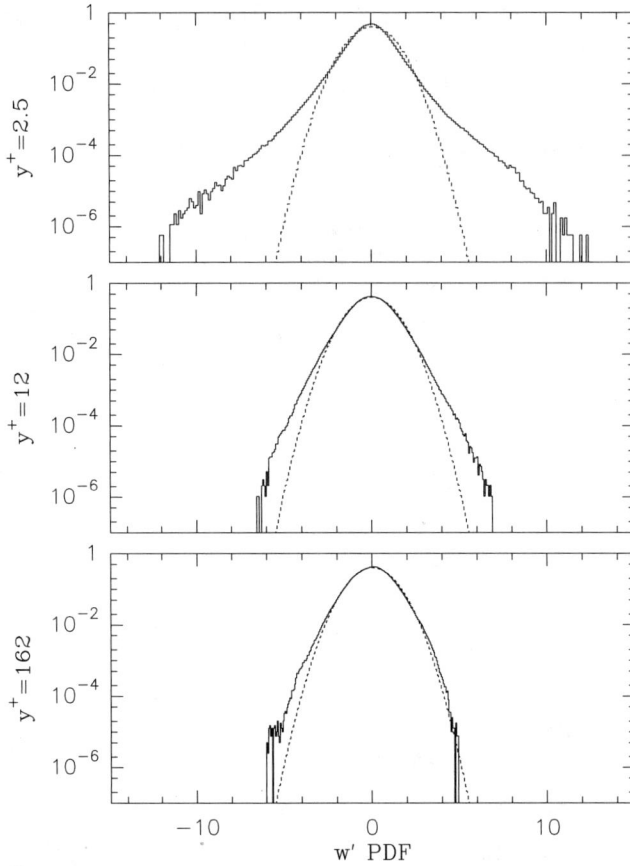

Figure 5.12. Channel spanwise velocity pdf at different distances from the wall. (Courtesy *Phys. Rev. E.* and E. Lamballais.)

other components v' and w' look like isotropic turbulence. The longitudinal velocity u' is now preferably negative, with excursions of low speeds. It seems then that the intermittent sweeps of high speed at the wall are balanced by low speeds in the channel center, which is quite natural if one thinks in terms of continuity. This coherent-vortex topology will be further discussed later in the case of weakly compressible channels and spatially growing boundary layers.

6 Current challenges for LES

We have clearly shown in the former chapters the advantages of the spectral eddy-viscosity models with, in particular, the possibility of accounting for local or semilocal effects in the neighborhood of the cutoff. More details on this point may be found in Sagaut [248], which contains many advanced aspects on LES modeling. However, in most industrial or environmental applications, the complexity of the computational domain prohibits the use of spectral methods. One thus has to deal with numerical codes written in physical space and employing finite-volume or finite-differences methods often with unstructured grids. This last point will not be considered in this book, although it is crucial for practical applications. We will present, however, simulations on orthogonal grids of mesh size varying in direction and location[1] and sometimes in curvilinear geometry. This chapter will mainly be devoted to models of the structure-function family with applications to isotropic turbulence, free-shear and separated flows, and boundary layers. We will also present in less detail alternative models such as the dynamic structure-function model, hyperviscosity model, mixed structure-function/hyperviscous model, and the mixed model.

6.1 Structure-function model

6.1.1 Formalism

The structure-function (SF) model is an attempt to go beyond the Smagorinsky model while keeping in physical space the same scalings as in spectral eddy-viscosity models. The original SF model is due to Métais and Lesieur [205]. It consists in building in physical space an eddy viscosity normalized by $\sqrt{E_{\bar{x}}(k_C)/k_C}$ with $k_C = \pi/\Delta x$. The spectrum $E_{\bar{x}}(k_C, t)$ is a local

---

[1] This was the case in particular for the channel presented earlier.

kinetic-energy spectrum at a given point $\vec{x}$, which has to be properly determined by assuming that turbulence is locally isotropic. This allows us to take
into account the spatial intermittency of turbulence. We first discard the peak
behavior[2] of $K(x)$ in Eq. (4.24) and then adjust the constant, as proposed by
Leslie and Quarini [177], by balancing, in a $k^{-5/3}$ inertial range extending
from zero to k_C, the subgrid-scale flux $2\int_0^{k_C} v_t k^2 \bar{E}(k)dk$ with the kinetic
energy flux ϵ. This yields

$$v_t(k_C) = \frac{2}{3} C_K^{-3/2} \left[\frac{E(k_C)}{k_C} \right]^{1/2}. \qquad (6.1)$$

We keep in mind that $E(k_C)$ is now a local kinetic-energy spectrum that has
to be evaluated in terms of physical-space quantities. The best candidate for
that is the second-order velocity structure function

$$F^{is}(r) = \left\langle [\vec{u}(\vec{x}, t) - \vec{u}(\vec{x} + \vec{r}, t)]^2 \right\rangle, \qquad (6.2)$$

where the label "is" stands for isotropic turbulence, and the brackets correspond to ensemble averaging. Indeed, we have already pointed out the equiva-
lence between Kolmogorov's $\epsilon^{2/3} k^{-5/3}$ spectrum and the $\langle \delta v(r)^2 \rangle \sim (\epsilon r)^{2/3}$
structure function derived from Eq. (1.42). We also recall Batchelor's relation
in isotropic turbulence

$$F^{is}(\Delta x) = 4 \int_0^\infty E(k, t) \left(1 - \frac{\sin(k\Delta x)}{k\Delta x} \right) dk. \qquad (6.3)$$

For the subgrid-modeling problem, we consider the following local structure
function:

$$F_2(\vec{x}, \Delta x) = \left\langle [\bar{\vec{u}}(\vec{x}, t) - \bar{\vec{u}}(\vec{x} + \vec{r}, t)]^2 \right\rangle_{\|\vec{r}\|=\Delta x}. \qquad (6.4)$$

The difference between (6.4) and relation (6.2) is that F_2 is calculated with
a local statistical average of square (filtered) velocity differences between $\vec{x}$
and the six closest points surrounding $\vec{x}$ on the computational grid. In some
cases, the average may be taken over four points parallel to a given plane.[3]
The equivalent Batchelor's formula is

$$F_2(\vec{x}, \Delta x) = 4 \int_0^{k_C} \bar{E}(k, t) \left(1 - \frac{\sin(k\Delta x)}{k\Delta x} \right) dk \qquad (6.5)$$

because the filtered field has no energy at modes larger than k_C. Assuming
again a $k^{-5/3}$ spectrum extending from zero to k_C, we obtain

$$v_t^{SF}(\vec{x}, \Delta x, t) = 0.105 \, C_K^{-3/2} \, \Delta x \, [F_2(\vec{x}, \Delta x)]^{1/2}. \qquad (6.6)$$

---

[2] Later we will show how to reintroduce the peak in terms of hyperviscosity.
[3] In a channel, for instance, the plane is parallel to the boundaries.

In fact, this derivation of the SF model equation is different from, and simpler than, the one proposed in the original paper of Métais and Lesieur [205] and may be found in Ducros [79].

6.1.2 Nonuniform grids

As emphasized in Lesieur ([170], p. 398), interpolations of Eq. (6.6) based on Kolmogorov's 2/3 law for the aforementioned structure function may be proposed if the computational grid is not regular (but still orthogonal). Let Δc be a mean mesh in the three spatial directions.[4] We have (in the six-point formulation)

$$F_2(\vec{x}, \Delta c) = \frac{1}{6} \sum_{i=1}^{3} F_2^{(i)} \left(\frac{\Delta c}{\Delta x_i} \right)^{2/3} \tag{6.7}$$

with

$$F_2^{(i)} = [\vec{u}(\vec{x}) - \vec{u}(\vec{x} + \Delta x_i\, \vec{e}_i)]^2 + [\vec{u}(\vec{x}) - \vec{u}(\vec{x} - \Delta x_i\, \vec{e}_i)]^2, \tag{6.8}$$

where $\vec{e}_i$ is the unit vector in the $\vec{x}_i$ direction.

6.1.3 Structure-function versus Smagorinsky models

We have found a relation between Smagorinsky's model and the structure-function model by replacing the velocity increments in the latter by first-order spatial derivatives. For the six-point formulation we get

$$\nu_t^{SF} \approx 0.777\,(C_S \Delta x)^2 \sqrt{2\bar{S}_{ij}\bar{S}_{ij} + \bar{\omega}_i\bar{\omega}_i}, \tag{6.9}$$

where $\vec{\bar{\omega}}$ is the vorticity of the filtered field and C_S is Smagorinsky's constant defined by Eq. (3.32) in terms of Kolmogorov's constant C_K. Then the SF model appears, within this crude first-order approximation, to be a combination of the Smagorinsky model in a strain and vortical version. Suppose as an example that we are in the stagnation regions between two quasi-two-dimensional vortices (e.g., in a mixing layer, a wake, or a round jet) when there is a low residual vorticity that is going to be stretched longitudinally. At this initial stage, and because vorticity in the stagnation region is low, the vortical term will be small compared with the strain term; thus, the SF model will be about 20% less dissipative than Smagorinsky's, which will favor the eventual stretching of longitudinal vortices.

Equation (6.9) does not clearly specify what happens for the two models within the vortices themselves. In fact, it can be reformulated by introducing

---

[4] It may be determined by a geometric mean, or of another type.

the Q quantity associated with resolved stresses to yield

$$v_t^{SF} \approx 1.01 \, (C_S \Delta x)^2 \sqrt{2\bar{S}_{ij}\bar{S}_{ij} + 2Q}. \tag{6.10}$$

Because we have seen in Chapter 2 that coherent vortices are well characterized by positive value of Q, the eddy viscosity given by Eq. (6.10) will be slightly greater than Smagorinsky's within the vortices. In "elliptic" regions between vortices, in contrast, where Q is negative, the structure-function eddy viscosity is smaller than Smagorinsky's, as previously noted.

6.1.4 Isotropic turbulence and free-shear flows

It has been shown in Métais and Lesieur ([205]; see also Lesieur [170], p. 398) that, for $C_K = 1.40$, the SF model gives a quite good $k^{-5/3}$ energy spectrum,[5] whereas the Smagorinsky model[6] exhibits more of a k^{-2} inertial range. These results confirm that the SF model performs better than Smagorinsky's model if nothing is done to reduce the natural (i.e., in developed turbulence) value of its constant. The SF model has also been applied to a spatially growing wake by Gonze [113] at zero molecular viscosity. The wake is initiated upstream by a constant velocity plus a deficit-velocity profile close to a top hat with a three-dimensionally isotropic, random perturbation regenerated at each time step. The numerical code used is the mixed spectral-compact code already presented for the channel computations.[7] A map of the vorticity modulus is presented in Figure 6.1. One clearly sees the shedding of alternate Karman vortices, which stretch intense longitudinal vortices in the stagnation regions between two opposite-sign vortices. In spatial mixing-layer LES using the same code with analogous upstream, random, three-dimensional perturbations of the same amplitude (see section 6.5), one sees the development of helical pairing between the large vortices. Nothing similar appears in our wake.

Let us mention the use of an eddy viscosity proportional to $\sqrt{E(k_C)/k_C}$, and hence not far from the SF model, by Sankaran and Menon [255] for LES of spray combustion in swirling jets. They were able to model kinematically and thermodynamically individual droplets of a dispersed spray in a gas-turbine combustor model. These researchers have provided impressive online animations of the jets and were able to study the effect of increased swirl, which enhances fuel–air mixing and favors combustion efficiency.

6.1.5 SF model, transition, and wall flows

It was at first quite a disappointment to realize that the SF model is, like Smagorinsky's, too dissipative for transition in a boundary layer forced

---

[5] The model has a nearly flat compensated $k^{5/3}E(k)$ spectrum.

[6] C_S is still given by Eq. (3.32).

[7] It is here the streamwise direction that is computed by finite differences of sixth order.

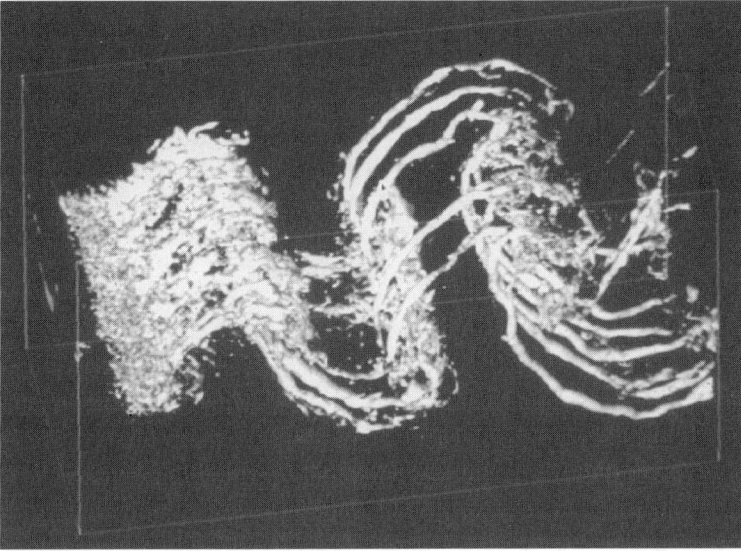

Figure 6.1. LES (using the SF model) of a spatially growing wake initiated upstream by a deficit profile close to a top-hat shape and perturbed by three-dimensional white noise. (Courtesy M.A. Gonze.)

upstream by a weak, three-dimensional, white-noise perturbation, for which it also yields relaminarization. Nor does it work very well in a channel. One might have thought that at least the four-point formulation in planes parallel to the wall would have eliminated the effect of the mean shear at the wall on the eddy viscosity. In fact, it turns out that the isotropic relation given by Eq. (6.5) introduces spurious inhomogeneous effects owing to large scales in the eddy viscosity, which increase the latter, and the SF model is too dissipative for quasi-two-dimensional or transitional situations. This is of course a real concern, especially for turbulent boundary layers or channel flows, and has motivated the development of two improved versions of the SF model: the selective structure-function model (SSF) and the filtered structure-function model (FSF), in which turbulence removes large-scale inhomogeneities before the SF model is applied.

6.2 Selective structure-function model

6.2.1 Formalism

In the SSF model of David [61], the eddy viscosity is switched off when the flow is not three-dimensional enough. We need therefore a criterion of three-dimensionalization, defined as follows: we consider at a given time the angle between the vorticity vector at a given grid point and the arithmetic mean of vorticity vectors at the six closest neighboring points (or the four closest points in the four-point formulation). If we carry out LES of isotropic turbulence at a resolution of 32^3–64^3, we find that the pdf peaks at an angle of 20°, which is thus the most probable value. Then, the eddy viscosity will be canceled at points where this angle is smaller than 20°. Compared with the original SF model, the SSF model dissipates the supergrid-scale energy at fewer points of the computational domain, and the model constant of 0.105 [see Eq. (6.6)] then has to be increased. David [61] chose to impose the same spatially averaged eddy viscosities for both the SF model and the SSF model; thus he obtained with the aid of isotropic test fields

$$\nu_t^{SSF}(\vec{x}, \Delta x, t) = 0.172 \; \Phi_{20°}(\vec{x}, t) \; C_K^{-3/2} \; \Delta x \; [F_2(\vec{x}, \Delta x)]^{1/2}, \quad (6.11)$$

where $\Phi_{20°}(\vec{x}, t)$ is a step function equal to zero at points of space at which the vorticity angle is smaller than 20° and equal to one if it is higher. We will give several applications of this model throughout the book. It allows good results to be obtained for various incompressible and compressible turbulent flows (see, for instance, Lesieur and Métais [168] for a review).

6.2.2 The problem of constant adjustment

However, and as noted by Ackerman and Métais [4], the SSF model presents some weaknesses that may render it difficult to adapt to very irregular meshes or to unstructured meshes. First, it seems obvious that the critical angle has to depend on the local numerical resolution. Indeed, for an infinitely refined resolution with a local grid size tending to zero, the angle tends to zero. Second, it is well known that the global kinetic-energy exchange between the supergrid and the subgrid scales for any eddy-viscosity model is given by $2\langle \nu_t \bar{S}_{ij} \bar{S}_{ij} \rangle$, where the operator $\langle \; \rangle$ is here a spatial average on the domain. If the cutoff wavenumber k_C is assumed to be located in a Kolmogorov cascade, the loss of energy by the supergrid scales is very close to the energy that cascades through the inertial subrange and is eventually dissipated by molecular viscosity in the dissipative range. So, from an energetic viewpoint, it would be more satisfying if the SSF model ensured the following relation:

$$2\langle \nu_t^{SSF} \bar{S}_{ij} \bar{S}_{ij} \rangle = 2\langle \nu_t^{SF} \bar{S}_{ij} \bar{S}_{ij} \rangle. \quad (6.12)$$

This leads to a modification of the constant arising in the SSF model, yielding improvements for the decay of isotropic turbulence. However, the classical SSF model still gives better results for wall flows.

6.3 Filtered structure-function model

The FSF model, developed by Ducros ([79]; see also Ducros et al. [81]) is described in Lesieur ([170], p. 399). The filtered field $\bar{u}_i$ is now submitted to a high-pass-filter $(\tilde{.})$ consisting of a Laplacian operator discretized by

second-order, centered, finite differences and iterated three times. We first
apply relation (6.5) to the high-pass-filtered field

$$\tilde{F}_2(\vec{x}, \Delta x) = 4 \int_0^{k_C} \tilde{E}(k) \left(1 - \frac{\sin(k\Delta x)}{k\Delta x}\right) dk, \qquad (6.13)$$

where $\tilde{F}_2(\vec{x}, \Delta x)$ is the second-order structure function of the high-pass-
filtered field $\widetilde{\tilde{u}}_i$, and $\tilde{E}(k)$ is its spectrum. This allows us (for isotropic turbulence) to relate $\tilde{F}_2$ to $\tilde{E}(k_C)$ and hence to $E(k_C)$ thanks to the transfer function
of the "tilde" operator determined with the aid of isotropic test fields. Using
Eq. (6.1) yields for the eddy viscosity

$$\nu_t^{\text{FSF}}(\vec{x}, \Delta x) = 0.0014 \, C_K^{-3/2} \, \Delta x \, [\tilde{F}_2(\vec{x}, \Delta x)]^{1/2}. \qquad (6.14)$$

A further advantage of the FSF model is that it does not contain adjustable
constants. We will show in the following sections very satisfactory applications of this model to mixing layers and to a boundary layer on a flat plate.

6.4 Temporal mixing layer

We present in Figure 6.2 a comparison between Smagorinsky's model, the
plain (nondynamic) spectral plateau-peak model, and the various structure-
function models (original, selective, and filtered versions). The comparison
is carried out in the case of a temporally growing mixing layer in a uniform-
density flow. As in the simulation of Figure 5.1, we use pseudospectral methods. For the wake, the molecular viscosity is zero, and we are in the realm
of Euler equation LES. We take a three-dimensional initial isotropic perturbation, but the domain now contains only two fundamental, longitudinal,
most-unstable wavelengths; thus, no helical pairing develops. Instead, we
see two big rollers oscillating in phase[8] along with stretching longitudinal
haipins, exactly as in the model of Bernal and Roshko [23], with very neat
alternate longitudinal vortices. Notice the strong resemblance among the results obtained with the plateau-peak, FSF, and SSF models. All give bigger spanwise and longitudinal vortices than the Smagorinsky and SF models
and exhibit considerably more small-scale variability. This confirms that both
modifications of the original SF model proceed in the right direction because the primary and secondary instabilities are less damped with the new
models. Note also that the SF model does not differ here very much from
Smagorinsky's.

---

[8] Such a configuration corresponds to "translative instability" from the work of Pierrehumbert
and Widnall [235] on secondary instabilities (Floquet-type analysis) of Stuart vortices.

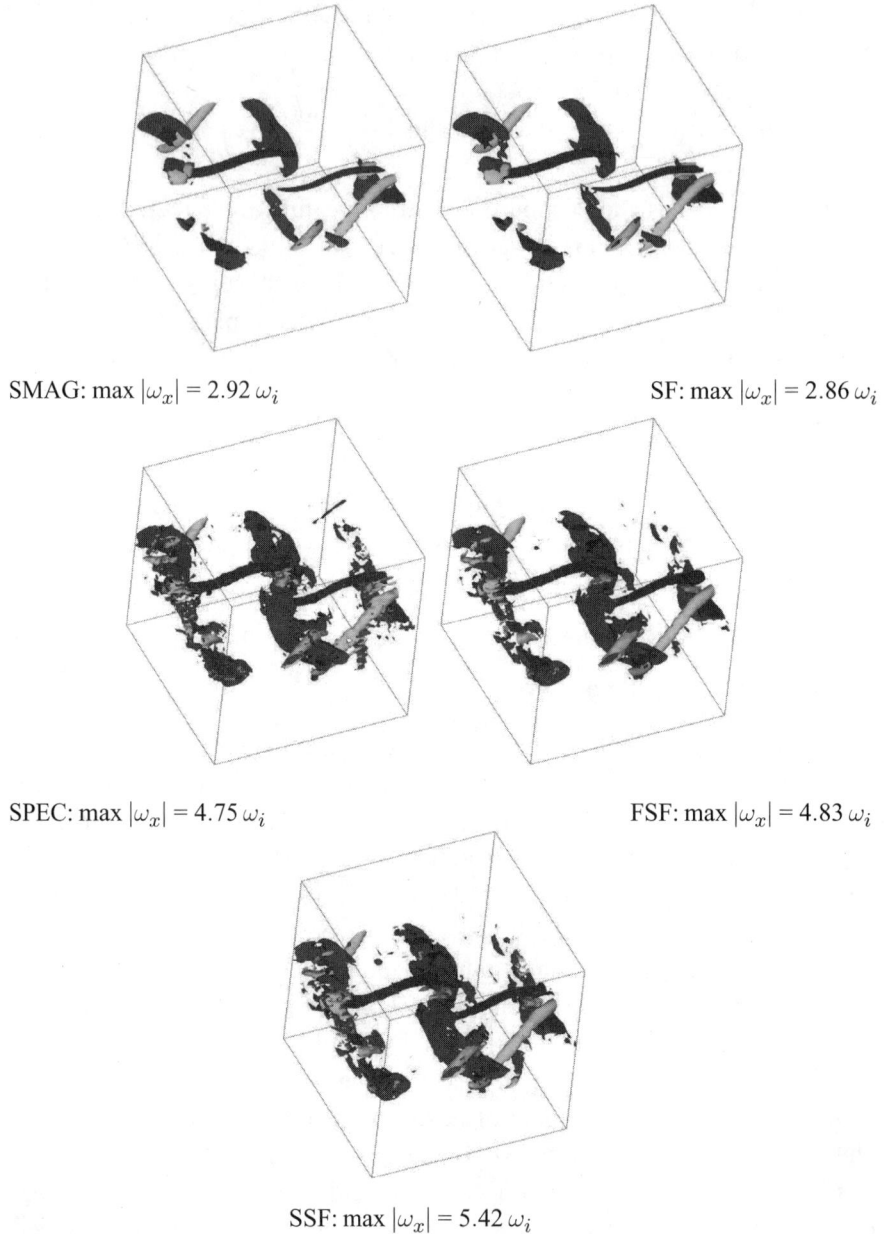

SMAG: max $|\omega_x| = 2.92\,\omega_i$ SF: max $|\omega_x| = 2.86\,\omega_i$

SPEC: max $|\omega_x| = 4.75\,\omega_i$ FSF: max $|\omega_x| = 4.83\,\omega_i$

SSF: max $|\omega_x| = 5.42\,\omega_i$

Figure 6.2. Comparison of various subgrid-scale models (Smagorinsky, SF, plateau-peak, FSF, SSF) applied to a temporal mixing layer visualized by isosurfaces $\omega_x = \omega_i$ (black), $\omega_x = -\omega_i$ (light gray), and $\omega_z = \omega_i = -2U/\delta_i$ (dark gray).

6.5 Spatial mixing layer

The temporal approximation is only a crude approximation of a spatially developing mixing layer, where one works in a frame traveling with the average velocity between the two layers. We consider now an incompressible mixing layer that is spatially developing between two streams of velocity U_1

Table 6.1. Table of the simulations presented here. L_x, L_y, and L_z denote the domain's dimensions in the streamwise, transverse, and spanwise directions, respectively. The corresponding numbers of collocation points N_x, N_y, and N_z are such that the meshes are cubic of side $\Delta \approx 0.29\, \delta_i$.

Run	ε_{2D}	ε_{3D}	Re	L_x/δ_i, L_y/δ_i, L_z/δ_i	N_x, N_y, N_z
DNSQ2D	10^{-3}	10^{-4}	100	140, 28, 14	480, 96, 48
FSFQ2D	10^{-4}	10^{-5}	∞	112, 28, 14	384, 96, 48
FSFQ2DW	10^{-4}	10^{-5}	∞	112, 28, 28	384, 96, 96
FSF3DW	0	10^{-4}	∞	112, 28, 28	384, 96, 96

and U_2 ($U_1 > U_2$). Further details can be found in Silvestrini [267]. The inflow is given by

$$\bar{u}(y) = \frac{U_1 + U_2}{2} + \frac{U_1 - U_2}{2}\, \tanh \frac{2y}{\delta_i}. \qquad (6.15)$$

The Reynolds number here is built on δ_i, and half the velocity difference $U = (U_1 - U_2)/2$. Characteristics of various runs are described in Table 6.1. We first compare a DNS at low Reynolds number ($Re = 100$) with a LES (without molecular viscosity) using the FSF model. Two small-amplitude random perturbations of Gaussian pdf are superimposed on this profile. The first is three-dimensional (i.e., a function of y, z, and t); its kinetic energy is denoted $\varepsilon_{3D}U^2$. The second (of energy $\varepsilon_{2D}U^2$) depends only on y and t. The ratio $\varepsilon_{2D}/\varepsilon_{3D}$ is set to 10, corresponding to quasi-two-dimensional perturbations. The DNS and the LES are henceforth referred to as DNSQ2D and FSFQ2D, respectively. It is important to notice that the low-Reynolds-number DNS requires more grid points than the LES, which will turn out to be much more turbulent. These simulations correspond to domains that are rather narrow in the spanwise direction. This is why the LESs have been redone by doubling the spanwise extent. These runs are either forced quasi-two-dimensionally (FSFQ2DW) or three-dimensionally (FSF3DW). The same mixed spectral-compact code already discussed for the channel and the wake is used here. Periodicity is assumed in the spanwise direction z. Sine and cosine expansions are used in the transverse direction y, enforcing free-slip boundary conditions. Nonreflective outflow boundary conditions are approximated by a multidimensional extension of Orlansky's discretization scheme with limiters on the phase velocity (see Gonze [113] for a detailed description of the numerical code).

Narrow domain

Figure 6.3 (top) shows an isosurface of the vorticity modulus. The vortex sheet undergoes oscillations leading to a first rollup further downstream. Subsequently, various pairings of KH vortices are observed. Again, thin, intense, longitudinal vortices are stretched as in Bernal and Roshko's [23] experiment. In run DNSQ2D, the vorticity magnitude during the run peaks at $2\omega_i$,

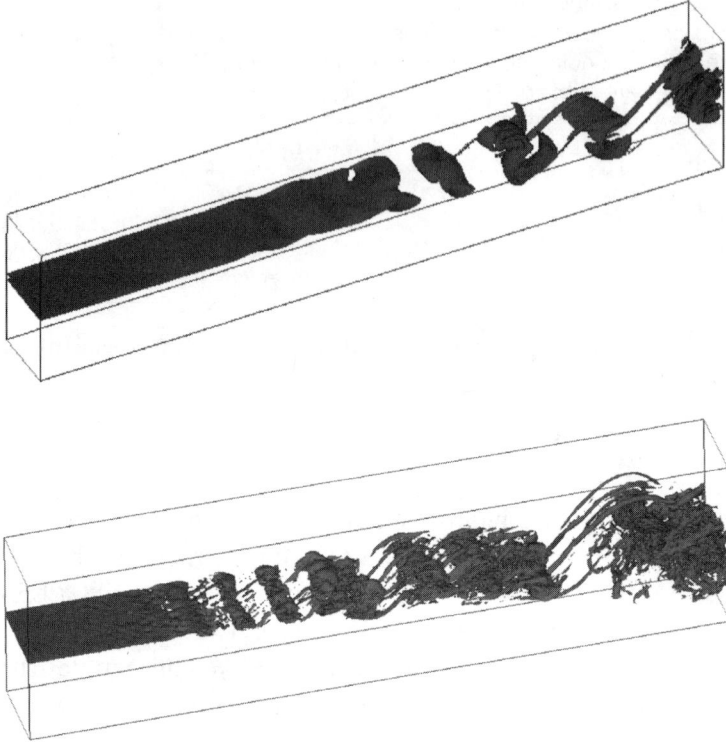

Figure 6.3. Perspective views of isovorticity surface. (Top) run DNSQ2D, $\|\bar{\omega}\| = \omega_i /3$; (bottom) run FSFQ2D, $\|\omega\| = (2/3)\omega_i$.

where $\omega_i = 2U/\delta_i$ is the maximal vorticity magnitude introduced at the inlet. Run FSFQ2D (Figure 6.3, bottom) is obviously much more turbulent than DNSQ2D, also exhibiting numerous oblique waves propagating along the upstream vortex sheet. The latter breaks down much faster, and the longitudinal vortices are stretched much more efficiently. Indeed, the maximal vorticity magnitude is now $\approx 4 \; \omega_i$ for the whole run. Rollup and pairing events occur much faster than in the DNS. Notice the complexity of the dynamics: we very clearly see three-dimensional waves propagating on the upstream vortex sheet before breaking into quasi-two-dimensional KH vortices stretching fine longitudinal vortices. There is a first pairing of KH vortices immediately followed by a tripling, leading to production of intense, small-scale, disorganized turbulence. Experimentally observed trends such as the doubling of the size and spanwise spacing of the longitudinal vortices at every pairing seem to be correctly reproduced. A movie corresponding to this LES simulation is on the CD-ROM (Animation 6-1). It is a calculation done on a parallel machine with sixteen processors using a domain-decomposition method in which the calculation volume is split into sixteen subdomains, each of which is associated with a processor. This is shown at the beginning of the movie with a fixed view of the vorticity norm. Afterward the calculation displays a domain duplicated by periodicity in the spanwise direction with vorticity modulus in light blue on the front and green on the bottom, positive longitudinal vorticity in red, and negative longitudinal vorticity in dark blue. The dynamics of longitudinal vortices is very complicated here with events corresponding to merging of same-sign vortices and events with apparent splitting of a vortex into two.

Wide domain

The vortical structure changes quite radically when the spanwise direction of the domain doubles. Figure 6.4, taken from Comte et al. [54], shows the low-pressure and vorticity fields developing from run FSF3DW. It is clear at least from the pressure that helical pairing develops, as in the experiments of Browand and Troutt [32]. When the forcing is a three-dimensional random white noise (run FSFQ2DW), helical pairing occurs again, as indicated by the low-pressure maps of Figure 6.5. It is, quite oddly, less intense than in the quasi-two-dimensional forcing case. However, none of these simulations has reached self-similarity, for rms velocity fluctuations have a departure of about 20% with respect to the experiments. Thus, calculations in longer domains are necessary to answer the important question about the exact topology (quasi two-dimensional + longitudinal hairpins versus helical-pairing) of coherent vortices in an unforced, constant-density, spatially growing, mixing layer.

6.6 Round jet

Our goal here is to demonstrate the ability of the LES to reproduce the coherent-vortex dynamics in the transitional region of a constant-density round jet properly. We also show the possibility of controlling the jet behavior by manipulating the inflow conditions. Some of the detailed results are presented in Urbin [284]. Because of the diversity of their coherent structures, axisymmetric jets constitute a prototype of free-shear flows of vital importance from both a fundamental as well as a more applied point of view. Indeed, a better understanding of the jet vortex structures should make possible the active control of the jet (e.g., spreading rate and mixing enhancement) for engineering applications (see, e.g., Zaman et al. [296]) especially in combustion and acoustics. In combustion, flames tend to follow vortical surfaces. In acoustics, it is clear that coherent vortices constitute, by the localized low pressures they induce, an important source of pressure waves and hence of noise. Manipulating the vortices in the jet of an airplane turboreactor will then have a direct effect on the noise emitted by either increasing or reducing it. Controlling a jet through forcings in the upstream conditions allows us to reduce the spreading of the jet in one direction and to enhance it in a perpendicular plane. This type

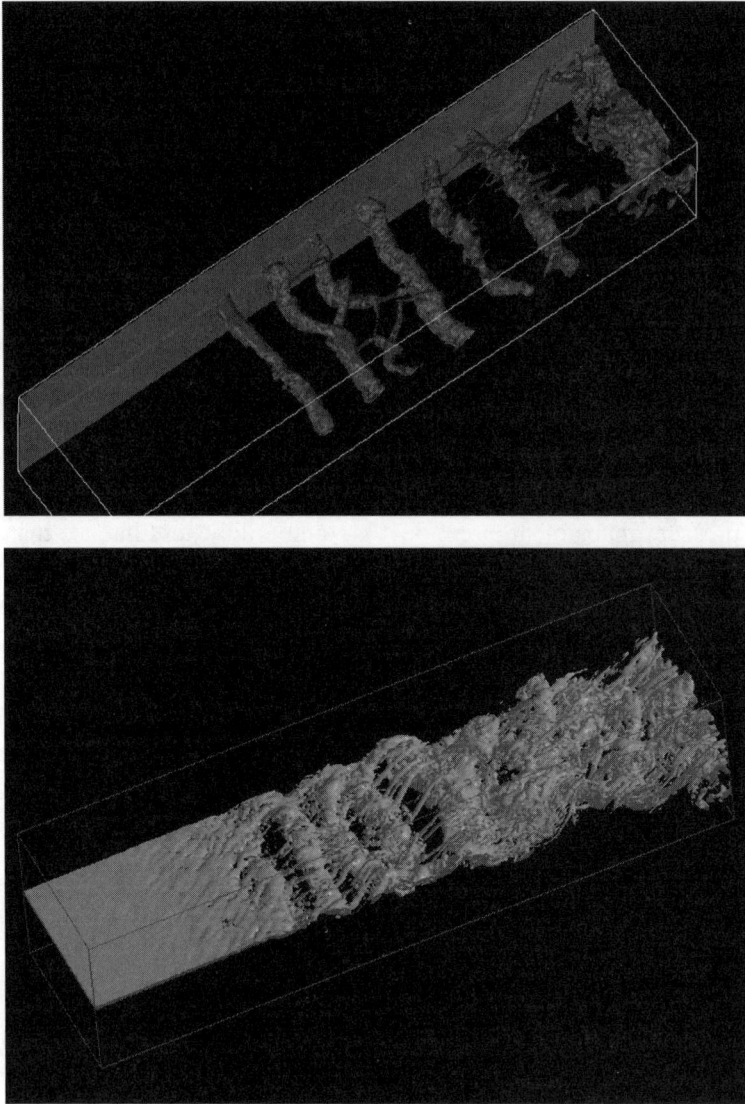

Figure 6.4. Wide domain for run FSF3DW. (Top) Low pressure; (bottom) high vorticity.

of control might be favorable for noise reduction. We recall that, in conditions corresponding to a supersonic transport plane taking off, the acoustic power emitted by the jet is proportional to some power (≈ 8) of the jet velocity. Jet spreading is used as a way to reduce velocity and hence jet noise.

In the past five years, important progress in the experimental methods of detection and identification has made possible a detailed investigation of the complex, three-dimensional, coherent vortices imbedded within this flow. For instance, the influence of the entrainment of the secondary streamwise vortices has been studied by Liepmann and Gharib [181]. On the numerical

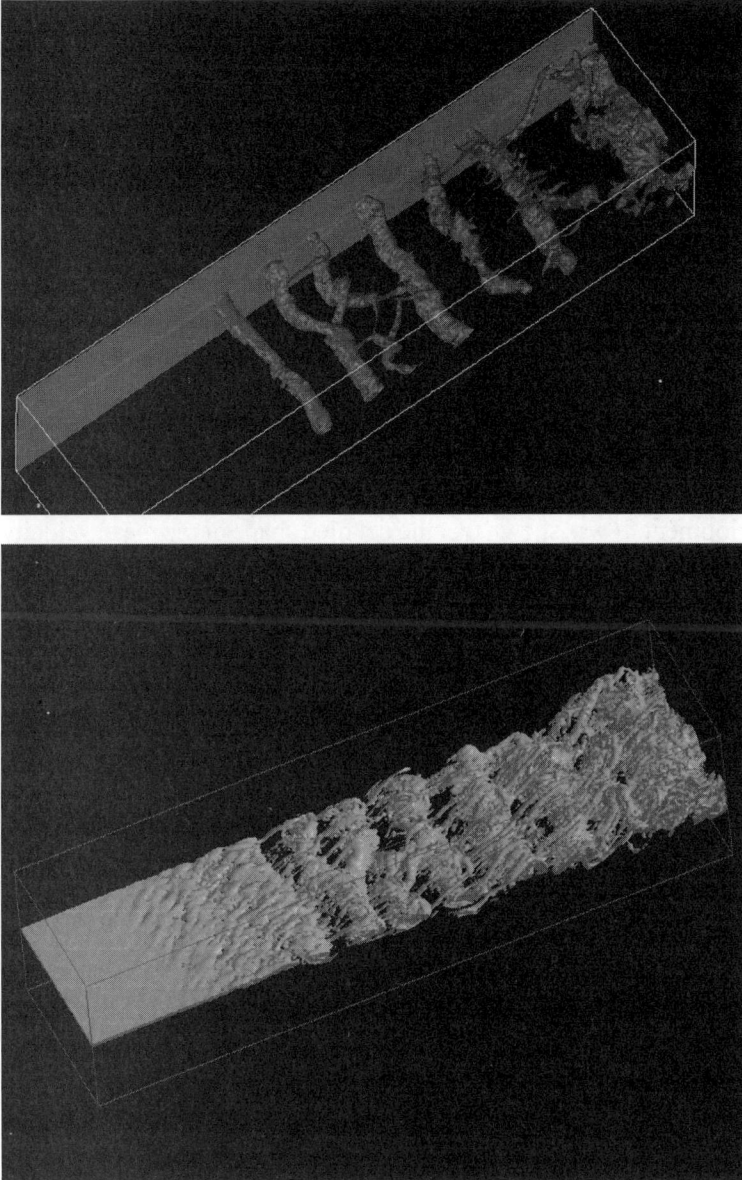

Figure 6.5. Wide domain for run FSFQ2DW. (Top) Low pressure; (bottom) high vorticity.

side, several simulations of two-dimensional or temporally evolving jets have been performed. However, very few have investigated the three-dimensionnal spatial development of the round jet. We show here how LES can be used to perform a statistical and topological numerical study of the spatial growth of the round jet from the nozzle up to sixteen diameters downstream.

We first present LES studies carried out by Urbin and co-workers [284–286] at a Reynolds number of 25,000. The subgrid model used here is the

SSF model, which is well adapted for transitional flows and accepts nonuniform grids. The LES filtered Navier–Stokes equations are solved using the TRIO-VF code. This industrial software was developed for thermal hydraulics applications by the Atomic Energy Commission in Grenoble. It uses a finite-volume method on a structured mesh. It has been used in many LESs of various flows, such as the study of Silveira-Neto et al. [266] for a backward-facing step, and is also a part of the backstep study presented in Chapter 2 and for which statistical results will be given in this chapter. This code, conservative from the point of view of momentum, is rather diffusive as far as kinetic energy is concerned. However, we will give new validations of the code, still for Chapter 2's backstep, in the following.

Experimental studies presented in Michalke and Hermann [209] have clearly pointed out the major effect of the inflow momentum boundary layer thickness θ and of the ratio R/θ (where R is the jet radius) on the jet's downstream development. It was shown that the detailed shape of the mean velocity profile strongly influences the nature of the coherent vortices appearing near the nozzle: Either axisymmetric structures (vortex rings) or helical structure can indeed develop.

In the simulations presented here, the flow inside the nozzle is not simulated, but a mean axial velocity profile in accordance with the experimental measurements is imposed:

$$U(r) = \frac{1}{2}U_0 \left[1 - \tanh\left(\frac{1}{4}\frac{R}{\theta}\left(\frac{r}{R} - \frac{R}{r} \right) \right) \right], \qquad (6.16)$$

where U_0 is the velocity on the axis. We restrict ourselves to a relatively small value of $R/\theta = 10$ because a correct resolution of the shear zone at the edge of the nozzle is crucial for reproducing the initial development of the instabilities. We consider a computational domain starting at the nozzle and extending to $16D$ (with $D = 2R$) downstream. The section perpendicular to the jet axis consists of a square of dimension $10D \times 10D$, which has been shown to be sufficient to avoid jet confinement. The computational mesh is refined at the jet shear layer (stretched mesh). The "natural" jet is forced upstream by the top-hat profile given by Eq. (6.16), on which a weak, three-dimensional, white noise is superposed. The "forced" jet development is controlled with the aid of various forms of deterministic inflow forcing (plus a white noise) designed to trigger specific types of three-dimensional coherent vortices.

6.6.1 The natural jet

The numerical approach has been validated by comparing the computed statistics with experimental results for the mean and for the rms fluctuating quantities. The frequency spectra have furthermore revealed at the end of the

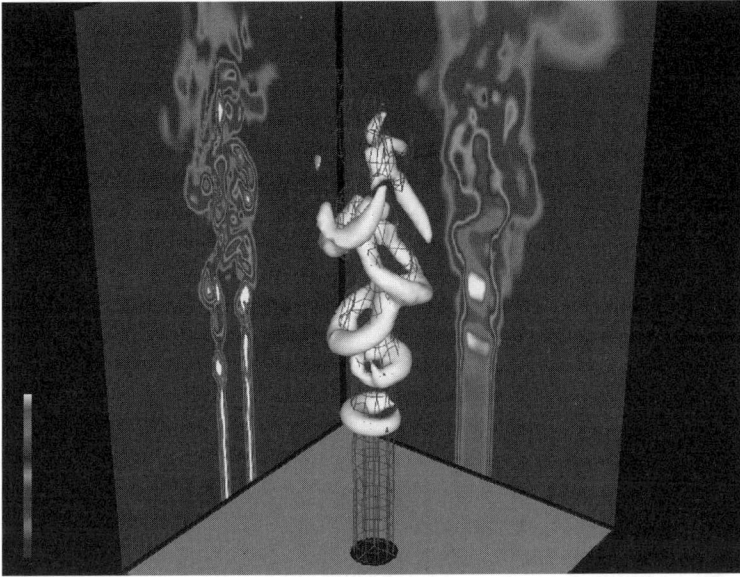

Figure 6.6. LES of the natural jet at a Reynolds number of 25,000 using the TRIO finite-volume code: instantaneous visualization. Light gray indicates a low-pressure isosurface wired isosurface; longitudinal velocity $U = U_0/2$. To the left is presented the xy cross section through the jet axis of the vorticity modulus, and to the right the x–z cross section of the velocity modulus. (Courtesy G. Urbin.)

potential core the emergence of a predominant, vortex-shedding, Strouhal number,[9] $Str_D = 0.35$, in good agreement with the experimental value.

A temporal linear-stability analysis performed on the inlet jet profile given by Eq. (6.16) (with $R/\theta = 10$) predicts a slightly higher amplification rate for the axisymmetric (varicose) mode than for the helical mode (see Michalke and Hermann [209]). In fact, LES verifies that the KH instability along the jet edge yields vortices further downstream that have mainly an axisymmetric toroidal shape. However, the simulations reveal that these vortices are not always present and intermitently bifurcate toward helically shaped vortices (see Urbin and Métais [285]).

Some of the following results are illustrated by Animation 6-2, which shows the same quantities as Figure 6.6 taken from the movie. The figure displays an original vortex arrangement subsequent to the varicose-mode growth, the "alternate pairing." Such a vortex interaction was previously observed by Fouillet [98] and Comte, et al. [53] in the DNS of a temporally evolving round jet at low Reynolds number ($Re = 2,000$). The direction normal to the symmetry plane of the toroidal vortices tends, during their advection downstream, to differ from the jet axis. The inclination angle of two consecutive vortices appears to be of opposite sign, eventually leading to a local pairing with an

---

[9] The Strouhal number is normalized by D and U_0.

alternate arrangement. Note that inclination of the vortex loops at the end of the potential core was experimentally observed by Petersen [234]. Experimental evidence of alternate pairing was also given by Broze and Hussain [34]. As explained by Silva and Métais [265], this alternate-pairing mode corresponds to the growth of a subharmonic perturbation (of wavelength double the one corresponding to the rings) developing after the formation of the primary rings. Indeed, let us consider a row of successive fundamental vortex rings that are perfectly axisymmetric whose axis is the jet axis [see Figure 6.7(a)] and whose spacing is L. Suppose they are submitted to a subharmonic interaction that entails displacing them respectively right and left with reference to the jet axis [see Figure 6.7(b)]. Each displaced vortex ring will feel a longitudinal velocity reduced at its exterior side with respect to the inner side, resulting in a torque tending to incline it as indicated on Figure 6.7(c). Finally, the edges of the rings that come close together will pair [see Figure 6.7(d)]. Alternate pairing therefore presents strong analogies with the helical-pairing mode observed in plane mixing layers (see the preceding discussion). The constant-density, free round jet at a Reynolds number of 25,000 has been recomputed by Silva [264] using the Grenoble spectral-compact code. Results are displayed in Animation 6-3. One finds qualitatively the same events as in Animation 6-2, but the flow is more chaotic and complex. This confirms that the TRIO finite-volume code is more diffusive than our spectral-compact code.

6.6.2 The forced jet

The previous natural jet simulations have therefore revealed three different types of vortical organization: the toroidal vortices (rings), the helical structure, and the alternate pairing. We now apply in the LES corresponding to Animation 6-2 a deterministic inflow perturbation to trigger one of these three particular flow organizations and to study the influence of the forcing on the statistics. Crow and Champagne [58] first noticed that the jet response is maximal with a preferred mode frequency corresponding to Str_D between 0.3 and 0.5.

One first applies a periodic fluctuation of frequency corresponding to $Str_D = 0.35$ superposed on the white noise.

Varicose excitation
We excite the varicose mode by imposing a periodic perturbation (alternately low speed and high speed) to the axial velocity at the nozzle:

$$U(r) + \epsilon\, U_0\, \sin\left(2\pi \frac{Str_D U_0}{D} t\right), \tag{6.17}$$

where $U(r)$ is given by Eq. (6.16) and $\epsilon = 1\%$.

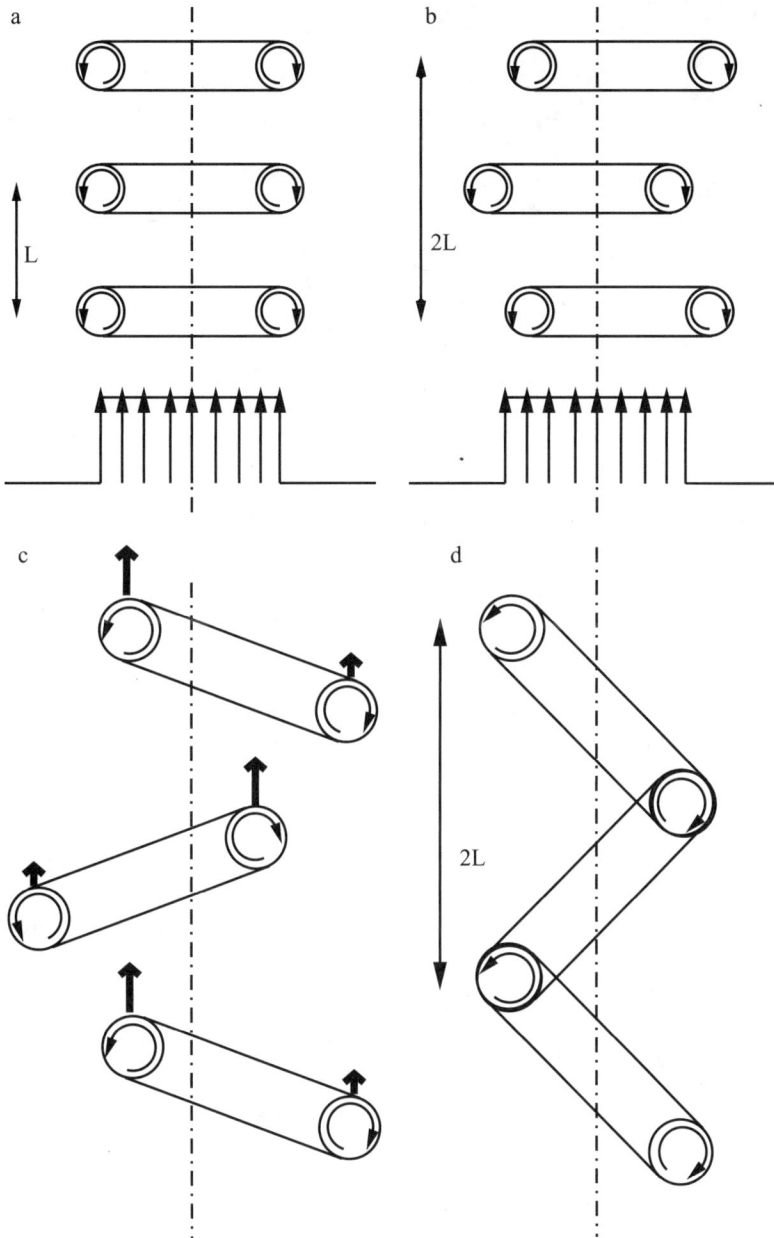

Figure 6.7. Schematic view of alternate pairing in a round jet. (Courtesy C. Silva and *Physics of Fluids* [265].)

Comparisons of the velocity fluctuations with experimental results show that, as opposed to the unexcited jet, a strong and fast amplification of the instability appears. Visualizations indicate that the varicose mode is now present at every instant from the beginning of the jet up to $x = 6D$. Vortex structures are more intense than in the natural case with well-marked and organized

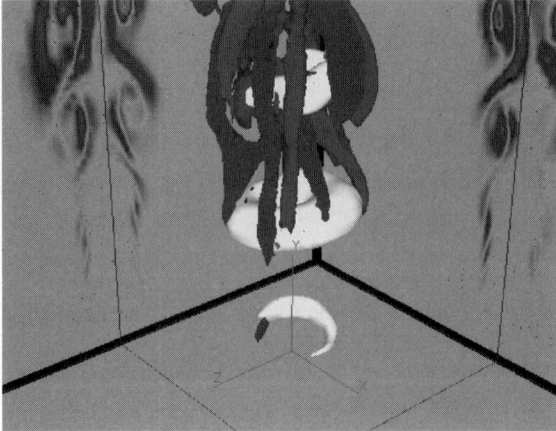

Figure 6.8. Forced jet at a Reynolds number of 25,000 using the TRIO finite-volume code with varicose-mode excitation. Black and gray indicate, respectively, positive and negative longitudinal vorticity isosurfaces corresponding to $\omega_x = \pm 1.2 U_0/D$. Shown are xy and xz cross sections (through the jet axis) of the longitudinal vorticity component ($min = -4U_0/D$; $max = +4U_0/D$).

pressure troughs (Figure 6.8).[10] Rings resulting from the varicose mode are now linked together by alternate longitudinal vortices, which have the same origin as the hairpins stretched between primary vortices that we have already observed in LES of quasi-two-dimensional mixing layers. The maximum vorticity within these structures in the jet simulation is about 40% of the vorticity of the associated rings. Longitudinal vortices of the same nature have already been observed experimentally at moderate Reynolds numbers (see, e.g., Lasheras et al. [157], Monkewitz and Pfizenmaier [213], and Liepmann and Gharib [181]). The present simulation would tend to indicate that they are also present at high Reynolds numbers.[11] These pairs of longitudinal vortices will, by velocity-induction mechanisms, entrain and eject fluid outside of the jet, thus creating transverse side jets and "branches." The latter were studied numerically in the temporal case by Martin and Meiburg [196] using vortex methods and by Abid and Brachet [3] using DNS. An interesting question concerns the azimuthal wavelength of the longitudinal vortices, which corresponds to the most unstable azimuthal mode of the primary rings in Widnall instability [292]. It is thus this wavelength that forces the wavelength of the stretched hairpins. The same phenomenon occurs experimentally in the turbulent mixing layer of Bernal and Roshko [23], where the spanwise wavelength of the longitudinal vortices[12] is the same as the most amplified translative mode of secondary instability analysis carried out by Pierrehumbert and Widnall [235] and previously mentioned.

---

[10] The axes on this figure and some of the following are misleading. We work actually with an orthonormal frame where x is the flow direction.

[11] One should, however, be cautious about the energy-diffusive character of the numerical code, which might artificially reduce the Reynolds number.

[12] It is of the order of two-thirds of the longitudinal wavelength of the spanwise vortices (see Lesieur [170], p. 85).

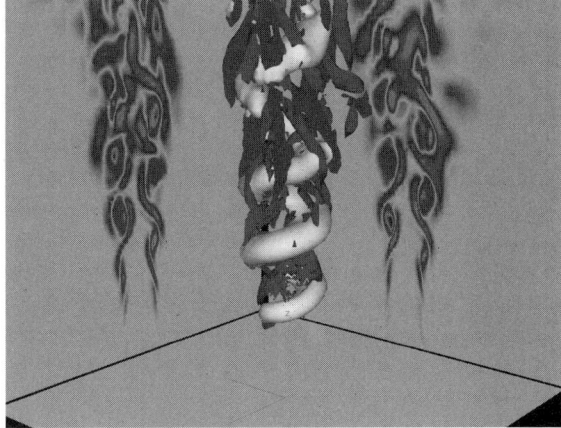

Figure 6.9. Forced jet at a Reynolds number of 25,000 using the TRIO finite-volume code with helical excitation. Quantities are the same as in Figure 6.8.

Helical excitation

The next excitation is designed to trigger the first helical mode by imposing the following inflow velocity profile:

$$U(r) + \epsilon \, U_0 \, \sin \left(\Theta - 2\pi \frac{Str_D U_0}{D} t \right) \frac{r}{D/2}, \qquad (6.18)$$

where Θ stands for the azimuthal angle. Indeed, the response of the jet consists of the development of a helical coherent vortex structure (Figure 6.9). This is in concordance with the results of Kusek et al. [150], who experimentally observed the helical mode development with an appropriate inflow excitation. The signature of the helical excitation on the statistics is an increase of the potential core length compared with the natural case and a reduction of the spreading rate. For the present jet (no swirl), and if molecular viscous effects are neglected, the Helmholtz–Kelvin theorem applies. Thus, the velocity circulation on a circular contour of large radius contained in a plane perpendicular to the jet axis and centered on the latter remains zero. This implies that the longitudinal vorticity flux through the surface limited by this contour is also zero. The present excitation gives rise to a helix structure that rotates in the anticlockwise direction when moving away from the nozzle: It is therefore associated with a negative longitudinal vorticity component. This generation of negative longitudinal vorticity has to be compensated by regions of positive longitudinal vorticity. Indeed, in the vicinity of the nozzle, Figure 6.9 shows the appearance of positive longitudinal vorticity on the helix edge. However, both positive and negative longitudinal vortices do appear further downstream, but the former are more intense than the latter. At $x = 4.5D$, the vorticity maximum within the positive vortices is $\approx 50\%$ of the vorticity (modulus) maximum within the helix, whereas it is only $\approx 25\%$ in the negative ones. Martin and Meiburg's [196] results display the same trend.

Flapping excitation

Following Silva and Métais [265], we define flapping excitation with the following upstream velocity:

$$U(r) + \epsilon\, U_0\, \sin\left(2\pi\, \frac{Str_D U_1}{D} t\right) \sin\left(\frac{2\pi\, y}{D}\right). \qquad (6.19)$$

Now half of the jet presents a speed excess, whereas a speed defect is imposed on the other half, and this excitation is applied alternately. Note that this perturbation has a preferred direction chosen along the y axis. The xy and xz planes are called, respectively, the bifurcating planes and bisecting planes. The resulting structures are analogous to those in Figure 6.6 except that the alternatively inclined vortex rings now appear from the nozzle. These inclined rings exhibit localized pairing and persist far downstream. As shown by Urbin [284] with the CEA TRIO-VF code and Silva and Métais [265] with the Grenoble spectral-compact code, the jet is squashed in the bisecting plane and widens in the bifurcating plane, where the spreading rate is strongly increased compared with the natural jet case. Many interesting DNS and LES studies of this type of jet have been carried out by Silva and Métais [265]. They show that taking the forcing Strouhal number equal to the jet harmonic mode Str_D is not the best choice for such effects. In fact, a varicose-flapping excitation may be introduced, which superposes a varicose forcing and a flapping forcing. Silva and Métais [265] show that the more dramatic effect is obtained in this case with a harmonic varicose frequency and a subharmonic flapping frequency. Their simulation is presented in Animation 6-4, where the jet is artificially rotating from the bisecting plane to the bifurcating one. This LES was done for a Reynolds number of 5,000, a resolution of $201 \times 128 \times 128$, a noise amplitude of 1%, and a forcing amplitude of 5%. Positive Q isosurfaces are shown.

We also recall studies on the "bifurcating" jet of Lee and Reynolds [161] and Reynolds et al. [241] (see also Parekh et al. [231]). They have experimentally shown that a properly combined axial and helical excitation can cause a turbulent round jet to split into two distinct jets. The experiment performed by Longmire and Duong [189] using a specially designed nozzle made of two half nozzles has displayed a similar vortex topology. One of the important technological applications of this peculiar excitation resides in the ability to polarize the jet in a preferential direction. Let us mention finally experiments carried out by Drobniak and Klajny [75] in an acoustically simulated jet at Reynolds numbers ranging from 5,000 to 100,000, where alternate pairing could be displayed.

We will show in Chapter 7 new results concerning the same type of control for the compressible jet in the subsonic and supersonic cases.

6.7 Backstep

We have already presented in Chapter 2 a study concerning the topology of coherent vortices generated over a backward-facing step in a flow of uniform density. It involves a LES based on the SSF model in its four-point formulation in planes parallel to the wall. No wall law is employed. The detailed computations can be found in Delcayre [69] and Dubief and Delcayre [78]. The code used, TRIO-VF, is the same as for the first jet simulations of Urbin [284] previously presented. We recall that the flow configuration is the same as for the DNS of Le et al. [159], which closely resembles the experiment performed by Jovic and Driver [138]. The calculations were performed with an inlet mean velocity profile obtained from Spalart's [273] boundary layer DNS at $Re_\theta = 670$ ($Re_{\delta_1} = 1,000$), where θ and δ_1 are the momentum and displacement thicknesses, respectively. For this particular profile, the boundary layer thickness is $\delta \approx 6.1\delta_1 = 1.2H$. The step-height Reynolds number is $Re_H = U_0 H/\nu \approx 5,100$. The inlet velocity profile is imposed at $0.3H$ (see Figure 2.9) upstream of the step with a three-dimensional random white noise of amplitude $\approx 1.25\%$ superimposed on the shear zone.

The computational domain extent is $15H$ downstream of the step, the spanwise direction size is $4H$, and in the vertical direction an expansion ratio of 1.2 is chosen, which corresponds to a domain height of $6H$. The total resolution of the computational domain is $97 \times 34 \times 46$. The grid is uniform in the spanwise direction and stretched in the direction normal to the walls to resolve the boundary layer. Because the grid is structured, the stretching of the upstream boundary layer also resolves the shear layer that develops downstream of the step. The spanwise grid spacing is constant and equal to 30 in wall units of the upstream turbulent boundary layer. The minimum streamwise resolution is $\Delta x_{\min}^+ = 11$ at the step. The maximum is at the exit boundary ($\Delta x_{\max}^+ = 70$). In the vertical direction, the minimum resolution is $\Delta y_{\min}^+ = 3.75$ (at the wall), and the maximum is $\Delta y_{\max}^+ = 110$.

The reattachment length X_R is overpredicted by the present LES, being $X_R = 7.2H$, which is far from the experimental measurement of Jovic and Driver ($X_R = 6.1H$). As already discussed in Chapter 2, this is attributable to the fact that the inflow boundary layer upstream of the step is not simulated deterministically (see Le et al. [159]). Indeed, we have seen in Chapter 2 that the absence of the turbulent longitudinal vortices associated with the inflow boundary layer induces a delay in the transition of the shear layer and an increase of the reattachment length. It was, however, shown by Westphal et al. [291] that, if the renormalized coordinate $X = (x - X_R)/X_R$ was used, statistical data were quite insensitive to the inflow conditions. This is confirmed by the comparison among the experiments of Jovic and Driver [138], the DNS of Le et al. [159], and the present LES presented at the reattachment point

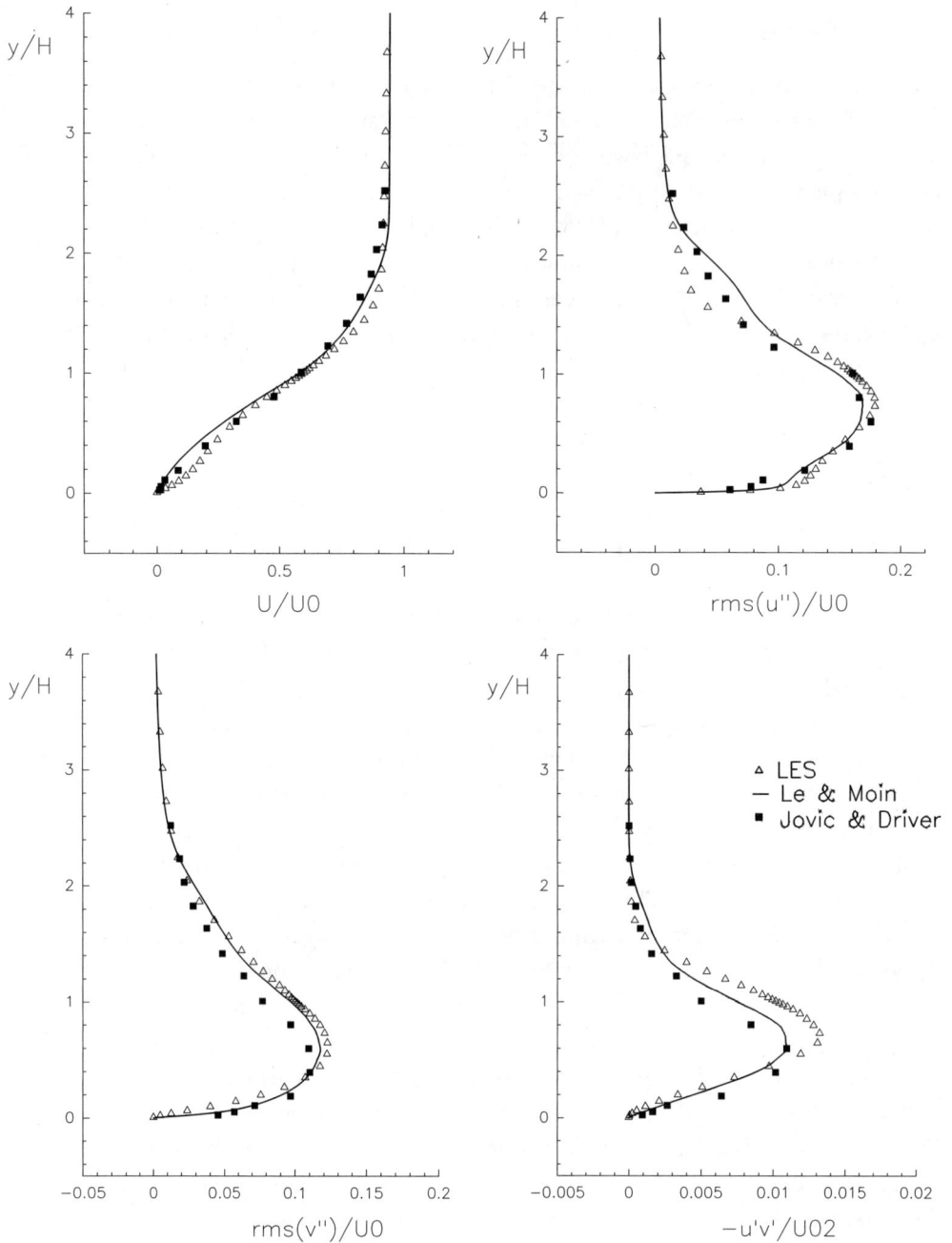

Figure 6.10. Backward-facing step. Shown are mean longitudinal velocity, rms longitudinal and vertical velocity, and Reynolds stresses at the reattachment point. The LES results are compared with the DNS results of Le et al. [159] and the experimental results of Jovic and Driver [138]. (Courtesy F. Delcayre.)

in Figure 6.10. Here, the statistical quantities are averaged over the spanwise direction, using the periodicity condition, and time. The agreement is found to be good.

Next, we carry out comparisons with the DNS for various values of the renormalized coordinate X (Figure 6.11). LES mean velocity profiles in the reattachment region almost collapse with the DNS. Both profiles still exhibit an inflectional point at the exit boundary ($X = 0.66$), indicating that the turbulent boundary layer is very far from being developed. This is explained by the topological study of Chapter 2, showing downstream of reattachment the persistance of big Λ vortices originating from the detached KH vortices formed downstream of the step. Longitudinal turbulent intensities also compare well with DNS, especially for $y/H < 1$. The slight underestimation of the longitudinal turbulent intensities for $y/H > 1$ in the LES could be attributable to the lack of longitudinal vortices in the inlet boundary layer flow. At the end of the domain, we can notice the development of a peak in the near-wall region. This peak proves the redevelopment of the boundary layer. Nevertheless, canonical turbulent boundary layer profiles are not yet recovered.

6.7.1 Strouhal numbers

Figure 6.12 shows the time-frequency pressure spectra at four positions in the flow: (1) just behind the step, (2) just before reattachment, (3) just behind reattachment, and (4) much farther downstream. Frequencies f are expressed in units of U_0/H and correspond to Strouhal numbers $Str = fH/U_0$. Position (1) is marked by a peak at $Str = 0.23$, indicating the shedding of KH vortices. At position (2), a second peak of higher amplitude is present at the subharmonic Strouhal number 0.12, corresponding physically to helical pairing. At positions (3) and (4), the two previous Strouhal numbers are still there, but a third peak forms at $Str = 0.07$, corresponding to the well-known flapping of the recirculation bubble. These different Strouhal numbers associated with the different unsteady phenomena are in good agreement with those previously found by other authors (see, e.g., Le et al. [159] and Arnal and Friedrich [9]). Such information regarding the pressure spectra and how they relate to the vortex dynamics is very important for acoustic studies and control of aerodynamic noise of cars, trains, and planes in particular.

6.8 New models

As already stressed, the use of Smagorinsky's model for the dynamic procedure is not compulsory, and other subgrid models may be candidates as well. We show here how the dynamic procedure may be implemented on the SF model.

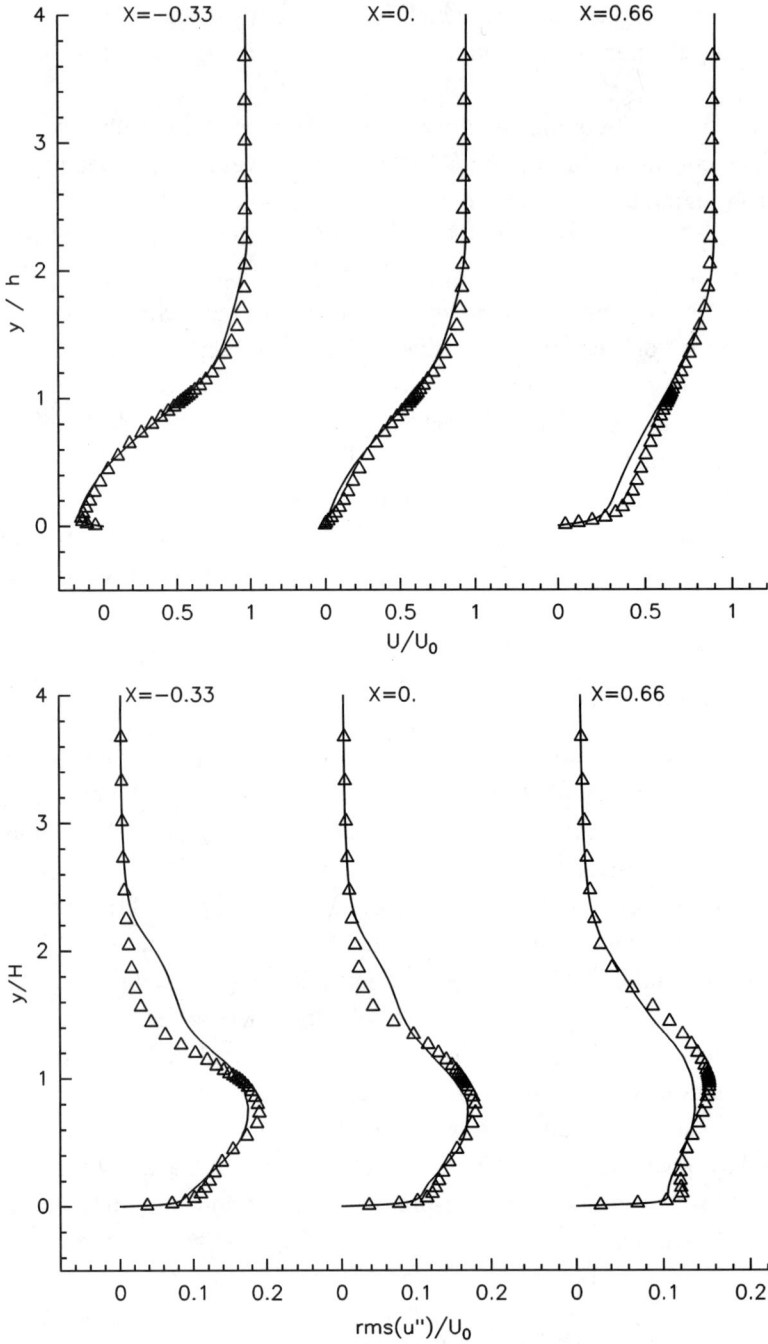

Figure 6.11. Backward-facing step. Shown are longitudinal mean and rms velocity profiles and a comparison of the LES with the DNS of Le et al. [159] (straight line). (Courtesy F. Delcayre.)

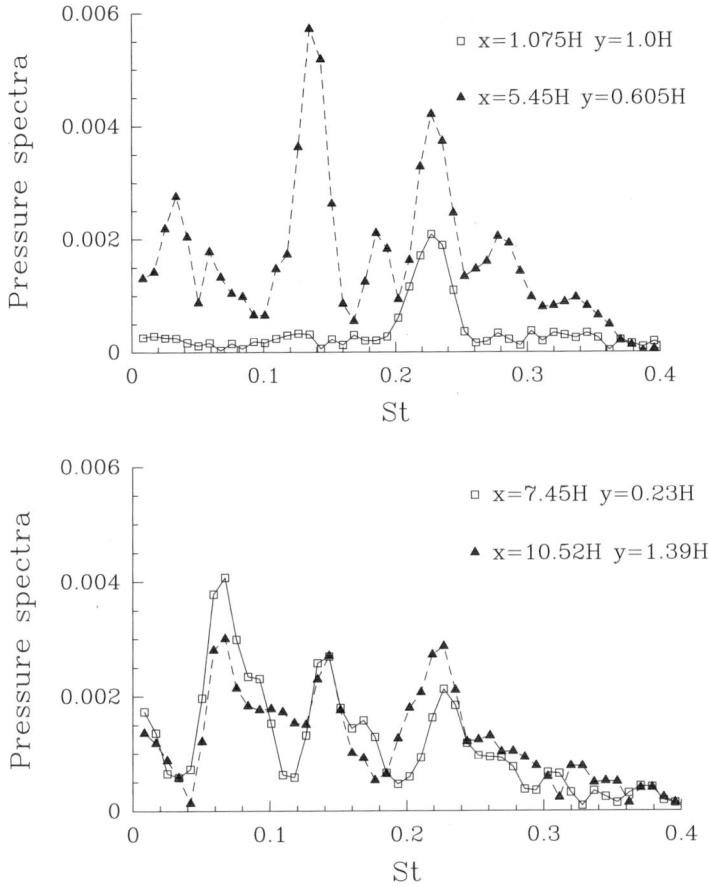

Figure 6.12. Backward-facing step. Shown are spanwise-averaged pressure frequency spectra at different positions obtained with LES. (Courtesy F. Delcayre.)

6.8.1 Dynamic structure-function model

We use the same notation as for Smagorinsky's dynamic model presented in Chapter 3, with a double filtering in physical space "$\bar{\cdot}$" and "$\tilde{\cdot}$" of respective width Δx and $\alpha \Delta x$, with $\alpha > 1$. We recall the notation

$$T_{ij} = \bar{u}_i \bar{u}_j - \overline{u_i u_j},$$

$$\mathcal{T}_{ij} = \widetilde{\bar{u}}_i \widetilde{\bar{u}}_j - \widetilde{\overline{u_i u_j}},$$

and

$$\mathcal{L}_{ij} = \widetilde{\bar{u}}_i \widetilde{\bar{u}}_j - \widetilde{\bar{u}_i \bar{u}_j}$$

with Germano's identity

$$\mathcal{L}_{ij} = \mathcal{T}_{ij} - \widetilde{T_{ij}}.$$

The unknown tensors will be modeled with the SF model eddy viscosity written as

$$\nu_t^{\text{SF}}(\vec{x}, \Delta x) = C_{\text{FS}}\, \Delta x\, [\bar{F}_2(\vec{x}, \Delta x)]^{1/2}, \tag{6.20}$$

where C_{FS} is the model constant to be determined, and $\bar{F}_2(\vec{x}, \Delta x)$ is the local second-order velocity structure function relative to the field $\bar{u}$. We can then easily show that

$$\widetilde{T}_{ij} - \frac{1}{3}\widetilde{T}_{ll}\, \delta_{ij} = 2C_{\text{FS}}\widetilde{\mathcal{A}_{ij}} \tag{6.21}$$

with

$$\mathcal{A}_{ij} = \Delta x\, [\bar{F}_2(\vec{x}, \Delta x)]^{1/2}\bar{S}_{ij}, \tag{6.22}$$

and

$$\mathcal{T}_{ij} - \frac{1}{3}\mathcal{T}_{ll}\, \delta_{ij} = 2C_{\text{FS}}\mathcal{B}_{ij} \tag{6.23}$$

with

$$\mathcal{B}_{ij} = \alpha\, \Delta x\, [\widetilde{\bar{F}}_2(\vec{x}, \alpha\, \Delta x)]^{1/2}\widetilde{\bar{S}}_{ij}. \tag{6.24}$$

From these we obtain

$$\mathcal{L}_{ij} - \frac{1}{3}\mathcal{L}_{ll}\, \delta_{ij} = 2C_{\text{FS}}M_{ij} \tag{6.25}$$

with the same relation as for Smagorinsky's dynamic mode

$$M_{ij} = \mathcal{B}_{ij} - \widetilde{\mathcal{A}_{ij}}. \tag{6.26}$$

We can then determine the constant C_{FS} with the same problems of overdetermination as for the Smagorinsky model.

As an example, El Hady and Zang [84] have applied the dynamic SF model to a compressible boundary above a long body.

6.8.2 Generalized hyperviscosities

One of the drawbacks of the SF model given by Eq. (6.6) is the absence of a cusp near k_{C}. However, EDQNM data show that the exponential form given in Eq. (4.26) of Chapter 4 can be correctly approximated by a power law of the type

$$\nu_t^*\left(\frac{k}{k_{\text{C}}}\right) = \left[1 + \nu_{tn}^*\left(\frac{k}{k_{\text{C}}}\right)^{2n}\right] \tag{6.27}$$

with $2n \approx 3.7$. Lesieur and Métais [168] have shown that ν_{tn}^* can be determined by considering the energy balance between explicit and subgrid-scale transfers. Indeed, let us write

$$\int_0^{k_C} 2\nu_t k^2 \bar{E}(k, t)\, dk = \epsilon \qquad (6.28)$$

with $\bar{E}(k, t) = C_K \epsilon^{2/3} k^{-5/3}$. This yields

$$\nu_{tn}^* = 0.512 \left(\frac{3n}{2} + 1 \right). \qquad (6.29)$$

In fact, the EDQNM value of $2n = 3.7$ is not so far from the exponent $2n = 4$ that would be obtained with a Laplacian operator iterated twice. Therefore, Lesieur and Métais [168] proposed the idea of a physical-space turbulent dissipative operator based on the SF model and with the "cusp" behavior taken into account. Among various possibilities, the following one seems to be the more simple for practical applications: The proposed subgrid dissipative operator is

$$2\frac{\partial}{\partial x_j} \left[0.661 \, \nu_t^{SF} \bar{S}_{ij} \right] + 0.661 \, \nu_{t2}^* \nu_t^{SF} \left(\frac{\Delta x}{\pi} \right)^4 (\nabla^2)^3 \bar{u}_i, \qquad (6.30)$$

where $\nu_{t2}^* = 2.048$ is given by Eq. (6.27) with $n = 2$. In Eq. (6.30), ν_t^{SF} is given by Eq. (6.6), and the constant 0.661 corresponds to the ratio of the plateau eddy viscosity ν_t^∞ given by Eq. (4.19) to the spectrally averaged eddy viscosity of Eq. (6.1). We have taken $k_C = \pi/\Delta x$. The expression (6.30) is interesting in the sense that it provides an eddy dissipation combining the SF model with a hyperviscosity $(\nabla^2)^3 \bar{u}_i$. The latter represents in physical space the action of the cusp in Kraichnan's spectral eddy viscosity. This model will be used in Chapter 8 for LES of the development of a baroclinic instability in a thermal front within a rotating stratified atmosphere.

6.8.3 Hyperviscosity

The model given by Eq. (6.30) contains a hyperviscous part. Hyperviscosity models are widely used in the study of geophysical flows because of their simplicity. They involve replacing the molecular dissipative operator $\nu\nabla^2$ by $(-1)^{\alpha-1}\nu_\alpha(\nabla^2)^\alpha$, where α is a positive integer. Unlike as in (6.30), ν_α is here a constant (positive) coefficient that has to be adjusted. This substitution has been widely used in two-dimensional isotropic turbulence (see Basdevant and Sadourny [15]) with values of α ranging from 2 to 8. Its effect is to shift the dissipation toward the neighborhood of k_c. This reduces the viscous dissipation of large coherent structures and is very useful for the study. In three-dimensional turbulence, hyperviscosity was used by Bartello et al. [14] to study the influence of a solid-body rotation in homogeneous turbulence with surprisingly good results. It was also used by Borue and Orszag [28] to check, in three-dimensional isotropic helical turbulence, the existence of a double $k^{-5/3}$ cascade of energy and helicity from large to small scales first discovered by André and Lesieur [6] via an EDQNM study.

6.8.4 Scale-similarity and mixed models

The lack of correlation between the subgrid-scale stress and the large-scale strain-rate tensors has led Bardina et al. [12] to propose an alternative subgrid-scale model called the scale-similarity model. This is based on a double-filtering approach and on the idea that the important interactions between the resolved and unresolved scales involve the smallest eddies of the former and the largest eddies of the latter. Bardina et al. suggest evaluating the subgrid tensor as

$$T_{ij} = \bar{\bar{u}}_i \bar{\bar{u}}_j - \overline{\overline{u_i \bar{u}_j}}. \tag{6.31}$$

The analysis of DNS and experimental data [12, 188] has shown that the modeled subgrid-scale stress deduced from Eq. (6.31) exhibits a good correlation with the real (measured) stress. However, when implemented in LES calculations, the model hardly dissipates any energy. It is therefore necessary to combine it with an eddy-viscosity-type model such as Smagorinsky's model to produce the "mixed" model. Along the lines of Bardina et al's. model, new formulations have been proposed to correct for this lack of dissipation. Liu et al. [188] have proposed the following model:

$$T_{ij} = C_{\rm L} \left(\tilde{\tilde{u}}_i \tilde{\tilde{u}}_j - \widetilde{\tilde{u}_i \tilde{u}_j} \right), \tag{6.32}$$

where $C_{\rm L}$ is a dimensionless coefficient. The operator $\tilde{\ }$ consists of a second filter of different size, as in the dynamic models in physical space already discussed.

Numerous developments on these types of models may be found in Sagaut [248].

6.8.5 Other approaches

Piomelli et al. [238] have studied an accelerated, spatially developing, turbulent boundary layer on a flat plate, using a dynamic Smagorinsky model in which the constant is calculated through a Lagrangian averaging procedure following the fluid parcel, as proposed by Meneveau et al. [201]. Their model eliminates sharp fluctuations of Smagorinsky's coefficient, which would otherwise destabilize the calculations, as already stressed. This method is particularly useful for studying coherent vortices, for we have seen that they tend to follow the flow. Piomelli et al.'s [238] LES studies show that high- and low-speed streaks are more elongated and less undulated, with a decrease of spanwise velocity fluctuations with respect to the streamwise ones. The number of quasi-longitudinal vortices is reduced, and they display the same trends as the velocity streaks. But the vorticity of these vortices is approximately unchanged with respect to the zero-pressure-gradient case. This is explained by Piomelli et al. [238] by the fact that additional vortex stretching caused by the acceleration is balanced by an increased dissipation of the vortices.

As stressed previously, the subgrid-scale tensor and scalar flux given by (3.18) are assumed to be strictly proportional to the grid-scale strain-rate tensor and buoyancy flux, respectively. Abbà et al. [1] have recently proposed an anisotropic formulation using eddy-viscosity and eddy-diffusivity tensors instead of scalar ones:

$$T_{ij} - \frac{1}{3}T_{ll}\,\delta_{ij} = 2\sum_{r,s} v^{t}_{ijrs}\bar{S}_{rs} - \frac{2}{3}\delta_{ij}\sum_{l,r,s,} v^{t}_{llrs}\bar{S}_{rs}. \qquad (6.33)$$

This formulation enables a better description of the small-scale anisotropy to be made. In conjunction with the dynamic procedure previously described, this model has been used successfully in LES of turbulent natural convection. A dynamic formulation has also been applied by the same authors to a round jet [2].

Let us quote also the so-called monotonically integrated large-eddy simulations (MILES) methods of Grinstein, Fureby, and co-workers. The method does not use any eddy viscosity but rather relies on numerical diffusion brought by high-resolution monotonic numerical algorithms used to solve Navier–Stokes equations. It is claimed that "implicit [subgrid-scale] models, provided by intrinsic nonlinear high-frequency filters built into the convection discretization, are coupled naturally to the resolvable scales of the flow" [104]. This approach is slightly controversial, and we prefer LES methods in which a physically relevant subgrid model is associated with the less possible diffusive numerical scheme (like pseudo-spectral methods). The MILES method does, however, give results as far as applications are concerned for free-shear flows and wall flows [104, 105, 115]. Simulations of a curved pipe carried out by Rütten et al. [246] belong to the same family.

Lastly, we mention that many people are currently working along the lines of defiltering procedures, which allow (if the filter is not a sharp filter in Fourier space) recovery of some of the subgrid stresses. The reader is referred to the review of Domaradzki and Adams [72] and the work of Geurts [110, 111] for more details.

6.9 Animations

Animation 6-1: LES of a constant-density mixing layer (no molecular viscosity). The first part presents a fixed view of the vorticity norm with splitting of the calculation volume into sixteen subdomains, each of which is associated to a computer processor. Afterward the domain is doubled by periodicity in the spanwise direction with vorticity modulus in light blue in the front (resp. green in the bottom) and positive longitudinal vorticity in red (resp. negative in dark blue). (Film 6-1.mpg; courtesy G. Silvestrini.)

Animation 6-2: LES of a constant-density, free round jet at a Reynolds number of 25,000 using the CEA TRIO-VF finite-volume code (see Figure 6.6 for details). The second part of the movie is in slow motion. (Film 6-2.mpg; courtesy G. Urbin.)

Animation 6-3: LES of a constant-density, free round jet at a Reynolds number of 25,000 using the LEGI spectral-compact code. Light gray and red indicate, respectively, low-pressure and positive Q isosurfaces. (Film 6-3.mpg; courtesy C. Silva.)

Animation 6-4: LES of a constant-density, forced round jet at a Reynolds number of 5,000 using the LEGI spectral-compact code, showing positive Q isosurfaces; the jet rotates from the bisecting to the bifurcating planes. It is forced by harmonic varicose and subharmonic flapping frequencies (Film 6-4.mpg; courtesy C. Silva.)

7 LES of compressible turbulence

Compressible turbulence has extremely important applications in subsonic, supersonic, and hypersonic aerodynamics. More generally, and even at low Mach numbers, strong density differences caused by intense heating (in combustion for instance) may have profound consequences on the flow structure and the associated mixing. Heating a wall may, for instance, completely destabilize a boundary layer, as will be shown for some applications in this chapter. The chapter is organized as follows. We will first present the compressible LES formalism for an ideal gas in a simple way, allowing us to generalize the use of incompressible subgrid models. This is possible using the concept of density-weighted Favre filtering together with the introduction of a macropressure and a macrotemperature related by the ideal-gas state equation. Then we will study compressible mixing layers at varying convective Mach numbers. Afterward we will consider low or moderate Mach numbers in boundary layers, channel, cavities, and separated flows and also a transonic rectangular cavity. A supersonic application relating to the European space shuttle *Hermès* rear-flap heating during atmospheric reentry will be discussed in detail. This problem, studied in Grenoble in 1993, has acquired a tragic topicality with the loss of the American *Columbia* shuttle on February 1, 2003. The latter disintegrated during reentry at an approximate elevation of 60 km and a speed of 21,000 km/h while making a turn at an angle of 57°. It seems that the left wing overheated, possibly because of damage to the protection tiles during takeoff.

Another aerospace application will be studied, namely, a heated channel related to the cooling of the European *Ariane V* launcher Vulcain engine.

7.1 Simplified compressible LES formalism

7.1.1 Compressible Navier–Stokes equations

In Cartesian coordinates, the compressible Navier–Stokes equations can be cast in the following flux form:

$$\frac{\partial U}{\partial t} + \frac{\partial F_1}{\partial x_1} + \frac{\partial F_2}{\partial x_2} + \frac{\partial F_3}{\partial x_3} = 0, \tag{7.1}$$

where

$$U = {}^T(\rho, \rho u_1, \rho u_2, \rho u_3, \rho e) \tag{7.2}$$

is a matrix of density, momentum, and total energy defined here for an *ideal gas*,

$$\rho e = \rho\, C_v\, T + \tfrac{1}{2}\rho(u_1^2 + u_2^2 + u_3^2), \tag{7.3}$$

so that gravity effects will be discarded. They are totally negligible in all the applications considered in the following. The fluxes F_i read $\forall i \in \{1, 2, 3\}$,

$$F_i = \begin{pmatrix} \rho u_i \\ \rho u_i u_1 - \sigma_{i1} \\ \rho u_i u_2 - \sigma_{i2} \\ \rho u_i u_3 - \sigma_{i3} \\ \rho e u_i - u_j \sigma_{ij} - \lambda \dfrac{\partial T}{\partial x_i} \end{pmatrix}, \tag{7.4}$$

where $\lambda = \rho C_p \kappa$ is the thermal conductivity (and κ is the thermal diffusivity). Equation (7.1) corresponds to continuity, momentum, and energy budgets.

The components σ_{ij} of the stress tensor are still given by Newton's law,

$$\sigma_{ij} = -p\, \delta_{ij} + 2\mu A_{ij}, \tag{7.5}$$

in which

$$A_{ij} = \frac{1}{2}\left[\frac{\partial u_j}{\partial x_i} + \frac{\partial u_i}{\partial x_j} - \frac{2}{3}(\nabla \cdot \boldsymbol{u})\delta_{ij}\right] \tag{7.6}$$

denotes now the deviator of the deformation tensor. This yields

$$F_i = \begin{pmatrix} \rho u_i \\ \rho u_i u_1 + p\, \delta_{i1} - 2\mu A_{i1} \\ \rho u_i u_2 + p\, \delta_{i2} - 2\mu A_{i2} \\ \rho u_i u_3 + p\, \delta_{i3} - 2\mu A_{i3} \\ (\rho e + p)u_i - 2\mu u_j A_{ij} - \lambda \dfrac{\partial T}{\partial x_i} \end{pmatrix}. \tag{7.7}$$

The Sutherland empirical law,

$$\mu(T) = \mu(273.15)\left(\frac{T}{273.15}\right)^{1/2} \frac{1 + S/273.15}{1 + S/T}, \tag{7.8}$$

with $\mu(273.15) = 1.711 \times 10^{-5}$ *Pl* and $S = 110.4$ K, and its extension to temperatures lower than 120 K,

$$\mu(T) = \mu(120)\, T/120 \quad \forall\, T < 120, \tag{7.9}$$

are prescribed for molecular viscosity. Thermal conductivity $\lambda(T)$ is related to the molecular Prandtl number by

$$Pr = \frac{\nu}{\kappa} = \frac{C_p \mu(T)}{\lambda(T)}. \tag{7.10}$$

In air at ambient temperature, the latter is equal to 0.7. The equation of state

$$p = R\rho T \tag{7.11}$$

closes the system, with $R = C_p - C_v = \frac{R}{M} = 287.06\,\mathrm{J \cdot kg^{-1} \cdot K^{-1}}$ for air at ambient temperature. We recall also that $\gamma = C_p/C_v$.

7.1.2 Compressible filtered equations

As in the incompressible regime, we introduce the low-pass filter *bar* of width Δx, which is larger by hypothesis than the Kolmogorov scale. Application of this filter[1] to the equations in the previous section yields

$$\frac{\partial \overline{U}}{\partial t} + \frac{\partial \overline{F}_1}{\partial x_1} + \frac{\partial \overline{F}_2}{\partial x_2} + \frac{\partial \overline{F}_3}{\partial x_3} = 0, \tag{7.12}$$

with

$$\overline{\rho e} = \overline{\rho c_v T} + \tfrac{1}{2}\overline{\rho(u_1^2 + u_2^2 + u_3^2)} \tag{7.13}$$

and

$$\overline{p} = \overline{\rho R T}. \tag{7.14}$$

Favre averages are density-weighted ensemble averages. As stressed by Favre [92], they were introduced by Hesselberg [123] and used by Dedebant and Wehrle [68], Van Mieghen and Dufour [287], and Blackadar [25] mainly to take into account density differences in meteorological turbulence. Favre [92, 93, 95] (see also [94]) developed the formalism extensively to study compressible turbulence. More specifically, the Favre average $\langle f \rangle_F$ of a function f is defined as

$$\langle f \rangle_F = \frac{\langle \rho f \rangle}{\langle \rho \rangle} \tag{7.15}$$

in such a way that we have

$$\langle \rho f \rangle = \langle \rho \rangle \langle f \rangle_F. \tag{7.16}$$

---

[1] The width is assumed to be the same in all three spatial directions.

Favre [92] thus studied turbulent-gas statistical equations, mass, momentum, kinetic energy (mean flow and fluctuations), enthalpy, entropy, and temperature. Favre averaging has been used all over the world with great success in compressible-turbulence modeling, and it is employed daily in the industry. Let us note also that the high-quality systematic measurements of turbulent velocity space–time correlations made by Favre, Coantic, Dumas, Dussauge, Gaviglio, and co-workers in Marseille are pioneering from the standpoint of coherent vortices (see Favre [95]). In particular, correlations for incompressible and supersonic boundary layers provide clear evidence that structures organized in space travel at some velocity in the flow. The reader is also referred to Smits and Dussauge [270] for very useful experimental information on compressible turbulence.

It is convenient for LES purposes to introduce a density-weighted Favre filter ($\tilde{\ }$) defined, for a given variable ϕ, by

$$\tilde{\phi} = \frac{\overline{\rho\phi}}{\overline{\rho}}. \tag{7.17}$$

This filter has of course nothing to do with the second filter introduced previously for the dynamic-model approach in physical space. We then have

$$\overline{\rho\phi} = \overline{\rho}\tilde{\phi} \tag{7.18}$$

and

$$\overline{U} = {}^{T}(\overline{\rho}, \overline{\rho}\tilde{u}_1, \overline{\rho}\tilde{u}_2, \overline{\rho}\tilde{u}_3, \overline{\rho}\,\tilde{e}). \tag{7.19}$$

The bar-filtered total energy can then be written as

$$\overline{\rho e} = \overline{\rho}\tilde{e} = \overline{\rho}\,C_v\,\tilde{T} + \tfrac{1}{2}\overline{\rho(u_1^2 + u_2^2 + u_3^2)} \tag{7.20}$$

if we neglect variations of C_v in the averaging. The same can be done for C_p and hence for R. The resolved bar-filtered fluxes $\overline{F_i}$ are then

$$\overline{F_i} = \begin{pmatrix} \overline{\rho}\tilde{u}_i \\ \overline{\rho u_i u_1} + \overline{p}\,\delta_{i1} - \overline{2\mu A_{i1}} \\ \overline{\rho u_i u_2} + \overline{p}\,\delta_{i2} - \overline{2\mu A_{i2}} \\ \overline{\rho u_i u_3} + \overline{p}\,\delta_{i3} - \overline{2\mu A_{i3}} \\ \overline{(\rho e + p)u_i} \quad - \overline{2\mu A_{ij}u_j} - \lambda\overline{\frac{\partial T}{\partial x_i}} \end{pmatrix} \tag{7.21}$$

with the filtered equation of state

$$\overline{p} = \overline{\rho}R\tilde{T}. \tag{7.22}$$

We reproduce now an analysis carried out in Grenoble by Ducros [79]. From the expression of $\overline{U}$ and the bar-filtered energy, it is clear that the relevant variables for the LES problem are $\overline{\rho}$ and the Favre-filtered quantities $\tilde{u}_i$ and $\tilde{e}$.

Once one has decided to work with these variables, it is then compulsory to introduce the subgrid-stress tensor $\overline{\overline{T}}$ of components

$$\mathcal{T}_{ij} = -\overline{\rho u_i u_j} + \overline{\rho} \tilde{u}_i \tilde{u}_j, \tag{7.23}$$

which we split into its isotropic and deviatoric parts, the latter being noted $\overline{\overline{\tau}}$:

$$\mathcal{T}_{ij} = \underbrace{\mathcal{T}_{ij} - \frac{1}{3} \mathcal{T}_{ll} \delta_{ij}}_{\tau_{ij}} + \frac{1}{3} \mathcal{T}_{ll} \delta_{ij}. \tag{7.24}$$

Equations (7.21) and (7.20) then read

$$\overline{F_i} = \begin{pmatrix} \overline{\rho} \tilde{u}_i \\ \overline{\rho} \tilde{u}_i \tilde{u}_1 + (\overline{p} - \frac{1}{3} \mathcal{T}_{ll}) \, \delta_{i1} - \tau_{i1} - \overline{2\mu A_{i1}} \\ \overline{\rho} \tilde{u}_i \tilde{u}_2 + (\overline{p} - \frac{1}{3} \mathcal{T}_{ll}) \, \delta_{i2} - \tau_{i2} - \overline{2\mu A_{i2}} \\ \overline{\rho} \tilde{u}_i \tilde{u}_3 + (\overline{p} - \frac{1}{3} \mathcal{T}_{ll}) \, \delta_{i3} - \tau_{i3} - \overline{2\mu A_{i3}} \\ \overline{(\rho e + p) u_i} \qquad\qquad - \overline{2\mu A_{ij} u_j} - \lambda \overline{\dfrac{\partial T}{\partial x_i}} \end{pmatrix} \tag{7.25}$$

and

$$\overline{\rho} \tilde{e} = \overline{\rho} \, C_v \tilde{T} + \frac{1}{2} \overline{\rho} \, (\tilde{u}_1{}^2 + \tilde{u}_2{}^2 + \tilde{u}_3{}^2) - \frac{1}{2} \mathcal{T}_{ll}, \tag{7.26}$$

where the latter is obtained by taking the trace of Eq. (7.23). In the same way as for the incompressible case (see Chapter 3), we introduce the *macropressure*

$$\varpi = \overline{p} - \frac{1}{3} \mathcal{T}_{ll}. \tag{7.27}$$

Let us rewrite Eq. (7.26) as

$$\overline{\rho} \tilde{e} = \overline{\rho} \, C_v \left(\tilde{T} - \frac{1}{2 C_v \overline{\rho}} \mathcal{T}_{ll} \right) + \frac{1}{2} \overline{\rho} \, (\tilde{u}_1{}^2 + \tilde{u}_2{}^2 + \tilde{u}_3{}^2) \tag{7.28}$$

and introduce a *macrotemperature*

$$\vartheta = \tilde{T} - \frac{1}{2 C_v \overline{\rho}} \mathcal{T}_{ll} \tag{7.29}$$

such that the bar-filter energy equation becomes

$$\overline{\rho} \tilde{e} = \overline{\rho} \, C_v \vartheta + \frac{1}{2} \overline{\rho} \, (\tilde{u}_1{}^2 + \tilde{u}_2{}^2 + \tilde{u}_3{}^2). \tag{7.30}$$

The bar-filtered equation of state (7.22) then reads

$$\begin{aligned} \varpi &= \overline{\rho} R \vartheta + \left(\frac{R}{2 C_v} - \frac{1}{3} \right) \mathcal{T}_{ll} \\ &= \overline{\rho} R \vartheta + \frac{3\gamma - 5}{6} \, \mathcal{T}_{ll}. \end{aligned} \tag{7.31}$$

If we consider from now on that ϖ is computable, it is sensible to involve it in the definition of a subgrid heat-flux vector, denoted $\boldsymbol{Q}$, of components

$$Q_i = -\overline{(\rho e + p)u_i} + (\overline{\rho}\tilde{e} + \varpi)\tilde{u}_i. \qquad (7.32)$$

The exact expression of the filtered fluxes is now

$$\overline{F_i} = \begin{pmatrix} \overline{\rho}\tilde{u}_i \\ \overline{\rho}\tilde{u}_i\tilde{u}_1 + \varpi\, \delta_{i1} - \tau_{i1} - \overline{2\mu A_{i1}} \\ \overline{\rho}\tilde{u}_i\tilde{u}_2 + \varpi\, \delta_{i2} - \tau_{i2} - \overline{2\mu A_{i2}} \\ \overline{\rho}\tilde{u}_i\tilde{u}_3 + \varpi\, \delta_{i3} - \tau_{i3} - \overline{2\mu A_{i3}} \\ (\overline{\rho}\tilde{e} + \varpi)\tilde{u}_i \quad - Q_i - \overline{2\mu A_{ij}u_j} - \lambda\dfrac{\overline{\partial T}}{\partial x_i} \end{pmatrix}. \qquad (7.33)$$

Now, we need an equation of state relating the macropressure and macrotemperature as well as closures to express the subgrid and molecular stresses.

7.1.3 Compressible LES equations

We introduce the subgrid Mach number

$$M_{\text{sgs}}^2 = \frac{\mathcal{T}_{ll}}{\overline{\rho}c^2}, \qquad (7.34)$$

where we recall that $c^2 = \gamma R T$, and thus we have also

$$\mathcal{T}_{ll} = \gamma M_{\text{sgs}}^2 \overline{p}. \qquad (7.35)$$

In compressible subgrid modeling, there are several options for the treatment of the uncomputable term $\mathcal{T}_{ll}$:

- Simply neglect it in front of $\overline{p}$, assuming as in Erlebacher et al. [87] that $\gamma M_{\text{sgs}}^2 \ll 1$ everywhere.
- Model it, as proposed by Yoshizawa [295], in a way that is consistent with the model chosen for $\overline{\overline{\tau}}$ (see, e.g., Moin et al. [211]). Note that this was the initial choice of Erlebacher et al. [86].
- Use the formalism of macropressure and macrotemperature we have introduced.

We choose the third possibility. If we consider Eq. (7.31), we notice that, for monatomic gases like argon or helium (for which $\gamma \approx 5/3$), the contribution of $\mathcal{T}_{ll}$ to this equation is quite negligible at any Mach number. It is extremely tempting to generalize this to air by assuming

$$\varpi \simeq \overline{\rho}R\vartheta, \qquad (7.36)$$

which is justified if the ratio of the second to the first term on the r.h.s. of Eq. (7.31) is much smaller than 1, which yields

$$\frac{|3\gamma - 5|}{6} \gamma M_{\text{sgs}}^2 \ll 1. \tag{7.37}$$

This is much better than assuming $\gamma M_{\text{sgs}}^2 \ll 1$: For instance, if $\gamma = 1.4$ (air at ambient temperature), the former condition is improved by a factor of 7.5; if $\gamma = 1.2$ (burnt gases), the factor is 4.3.

Notice that the same modified temperature as our macrotemperature is present in the work of Vreman et al. [289]. However, there it is associated with a different pressure than the macropressure, namely,

$$p_v = \overline{p} - \frac{\gamma - 1}{2} T_{ll}, \tag{7.38}$$

which is exactly $\overline{\rho} R \vartheta$. Their analysis of various energy subgrid terms differs from that of Ducros [79].

Now we return to Eq. (7.33), for which we invoke the usual eddy-viscosity and diffusivity models in terms of Favre-filtered quantities in the form

$$\tau_{ij} \simeq \overline{\rho} \nu_t \widetilde{A}_{ij}, \tag{7.39}$$

$$Q_i \simeq \overline{\rho} \, C_p \, \frac{\nu_t}{Pr_t} \frac{\partial \vartheta}{\partial x_i}. \tag{7.40}$$

Expressions for $\nu_t(\widetilde{u})$ and Pr_t used in the following compressible simulations correspond to the incompressible models.

The remaining noncomputable terms are molecular viscous and diffusive terms, which can be considered of less importance at sufficiently large Reynolds number.[2] We therefore simply replace Eq. (7.33) by

$$\overline{F_i} \simeq \begin{pmatrix} \overline{\rho}\widetilde{u}_i \\ \overline{\rho}\widetilde{u}_i\widetilde{u}_1 + \varpi \, \delta_{i1} - 2(\bar{\mu} + \bar{\rho}\nu_t)\widetilde{A}_{i1} \\ \overline{\rho}\widetilde{u}_i\widetilde{u}_2 + \varpi \, \delta_{i2} - 2(\bar{\mu} + \bar{\rho}\nu_t)\widetilde{A}_{i2} \\ \overline{\rho}\widetilde{u}_i\widetilde{u}_3 + \varpi \, \delta_{i3} - 2(\bar{\mu} + \bar{\rho}\nu_t)\widetilde{A}_{i3} \\ (\overline{\rho}\widetilde{e} + \varpi)\widetilde{u}_i \quad - 2\bar{\mu}\widetilde{A}_{ij}\widetilde{u}_j - \left[\bar{\lambda} + \bar{\rho}C_p\frac{\nu_t}{Pr_t}\right]\frac{\partial \vartheta}{\partial x_i} \end{pmatrix}, \tag{7.41}$$

in which $\bar{\mu}$ and $\bar{\lambda}$ are linked to ϑ through the Sutherland relation (7.8) with a constant molecular Prandtl number assumption $Pr = C_p\bar{\mu}(\vartheta)/\bar{\lambda}(\vartheta) = 0.7$ being made.

What is remarkable is that this system is equivalent to the compressible Navier–Stokes equations with the following changes: $u_i \rightarrow \widetilde{u}_i$, $\rho \rightarrow \bar{\rho}$,

---

[2] This is certainly questionable in a hypersonic boundary layer close to the wall, where the intense heating significantly increases molecular diffusion. However, the LES of the temporal boundary layer at Mach 4.5 carried out successfully by Ducros et al. [80] using the SF model does not seem to be affected by major problems at the wall.

$T \to \vartheta, p \to \varpi, e \to \tilde{e}, \mu \to \bar{\mu} + \bar{\rho}\nu_t$ (except in the energy equation), and $\lambda \to \bar{\lambda} + \bar{\rho}C_p\nu_t/Pr_t$. Such a choice had been made heuristically in Grenoble before the present formalism was developed (see, e.g., Normand and Lesieur [221]). The heuristic choice was based on arguments of the type that small scales could not be significantly affected by compressibility if the large-scale Mach number was not too high. In fact, we have seen that the introduction of macropressure and macrotemperature is a powerful tool, allowing us to push the validity of this model to much higher Mach numbers.

7.2 Compressible mixing layer

7.2.1 Convective Mach number

We first consider a spatially growing mixing layer between two parallel flows in the same direction of velocities $\bar{U}_1$ and $\bar{U}_2$. We assume that the two flows have different densities ρ_1 and ρ_2 at infinity in the transverse direction but maintain the same pressure. Let T_1 and T_2 be the corresponding temperatures. The associated sound speeds are $c_1 = \sqrt{\gamma R T_1}$ and $c_2 = \sqrt{\gamma R T_2}$ (we suppose that γ and R are the same in each layer). In fact, the interesting Mach numbers are the convective Mach numbers of the two layers, $M_c^{(1)}$ and $M_c^{(2)}$, in a frame moving with U_c, the velocity of the large vortices (Bogdanoff [27]). These Mach numbers are

$$M_c^{(1)} = \frac{U_1 - U_c}{c_1}, \quad M_c^{(2)} = \frac{U_c - U_2}{c_2}. \tag{7.42}$$

Assuming continuity of the dynamic pressure in the two flows about the stagnation point between vortices, we can show that

$$U_c = \frac{U_1 c_2 + U_2 c_1}{c_1 + c_2} = U_1 \frac{1 + \frac{U_2}{U_1}\sqrt{\rho_2/\rho_1}}{1 + \sqrt{\rho_2/\rho_1}}. \tag{7.43}$$

Then, within this assumption, the convective Mach numbers are both equal to

$$M_c = \frac{U_1 - U_2}{c_1 + c_2} = \frac{U}{\bar{c}}, \tag{7.44}$$

where $2U$ is the velocity difference, and $\bar{c} = (c_1 + c_2)/2$ is an average sound velocity between the two layers. This expression allows us to recover the value $U_c = (U_1 + U_2)/2$ in the incompressible uniform-density case. Papamoschou and Roshko [230] have shown experimentally that the equality of the two convective Mach numbers is valid up to $M_c \approx 0.6$, which means that the dynamic pressure continuity assumption made at the stagnation region no longer holds above this threshold. These experiments also show a dramatic decrease of the spreading rate of the mixing layer with respect to the incompressible value between $M_c \approx 0.4$ and $M_c \approx 0.9$. In fact, above the value of 0.6, what we call

M_c is the highest of the two convective Mach numbers. Above $M_c = 1$, the spreading rate saturates at about 40% the value of the incompressible case.

7.2.2 Temporal mixing layer

For a temporal mixing layer, the previous relations apply with $U_1 = -U_2 = U$ and thus the convective Mach number is still given by

$$M_c = \frac{U}{\bar{c}}. \tag{7.45}$$

Because $\rho_2 \neq \rho_1$, the velocity of KH vortices is no longer zero as in the constant-density case and now equals $U(1 - \sqrt{\rho_2/\rho_1})/(1 + \sqrt{\rho_2/\rho_1})$. The initial convective Mach number $M_c^{(i)}$ will then be defined by Eq. (7.45). If it is greater than 0.6, it is probable that the convective Mach numbers of both layers will become different from $M_c^{(i)}$.

We first review the linear-stability analyses of the compressible temporal mixing layer, following Lesieur ([170], p. 442). The inviscid linear-stability analysis in the two-dimensional case was performed by Lessen et al. [178, 179] and Blumen [26]. The stability diagram found by the latter (for $\gamma = 1.4$) shows that the maximum amplification rate is a decreasing function of the initial Mach number with a drastic change in the slope at $M_c^{(i)} = 0.6$. Two-dimensional DNS of Normand [220] show an inhibition of KH instability for $M_c^{(i)} > 0.6$: There is hardly any rollup of the vortices, which remain extremely flat and merge "longitudinally" without turning around each other. In contrast, for $M_c^{(i)} \leq 0.6$, rollup and pairing occur qualitatively in the same fashion as in the incompressible case, although they are delayed by factors corresponding exactly to the amplification rates predicted by Blumen [26]. Another interesting feature in these two-dimensional simulations is the appearance of shocklets on the edge of the vortices above $M_c^{(i)} = 0.7$–0.8. They are displayed in Lesieur ([170], p. 443) and are analogous to shocks on a transonic wing. It was shown numerically by Fouillet [98] that they disappear in two dimensions at higher convective Mach numbers.

The three-dimensional linear-stability analysis of the compressible temporal mixing layer was carried out by Sandham and Reynolds [253, 254]. They showed that the dominant instability becomes three-dimensional when $M_c^{(i)}$ exceeds 0.6. They also carried out DNS, with the initial forcing yielding KH vortices undergoing translative instability without compressibility, and they found at a convective Mach number of 1 a set of large staggered Λ vortices. Fouillet [98] (see also Comte et al. [53]) carried out a DNS of a compressible temporal mixing layer using the compressible Navier–Stokes solver COMPRESS developed in Grenoble by Normand and Lesieur [221]. It uses MacCormack's [193] predictor–corrector scheme of fourth order (for nonlinear terms) in space, as modified by Gottlieb and Turkel [114]. Time accuracy is of second order. The Reynolds number, based on U and the initial vorticity thickness, is 100, as in the incompressible DNS of Comte et al. [52]. Fouillet also takes the same domain size in the x direction, L_x, equal to four fundamental Michalke's [208] incompressible waves. Such a system, perturbed by a three-dimensional white-noise initial perturbation superposed to the hyperbolic-tangent velocity, was able to produce the helical-pairing interaction in the incompressible DNS of Comte et al. [52]. Recall that we have also recovered such an interaction using spectral eddy-viscosity LES in Chapter 5. Fouillet's initial temperature profile is determined with the aid of Crocco–Busemann's relation, and he still applies the three-dimensional random perturbation. The spanwise size of the domain L_z verifies the relation

$$M_c^{(i)} \cos \theta = 0.6, \qquad (7.46)$$

with $\tan \theta = L_z/L_x$. Such a relation comes from the linear-stability study of Sandham and Reynolds [253] and enables L_z to be fixed to the most amplified spanwise wavelength within the linear-stability study. The resolution of the simulation is 48^3 with a grid refinement in the central vortical region. We describe now DNSs at three different initial convective Mach numbers: 0.3, 0.8, and 1. Figure 7.1 shows, for $M_c^{(i)} = 0.3$, the time evolution at three instants $(25, 45, \text{ and } 50 \, \delta_i/U$, from top to bottom) of a top view of low-pressure isosurfaces (left) and vortex lines colored by pressure (right). One sees very nicely the staggered organization of the vortex lines yielding the helical pairing. This is very close to what we have observed in incompressible DNS and LES with the same forcing.

Figure 7.2 shows the same simulation (instants 50, 60, and 100) for $M_c^{(i)} = 0.8$. One still sees the formation of a staggered pattern at times delayed by compressibility, but there is no helical pairing at the end, where the big Λs lie above each other, as a side view indicates. An interesting feature is also the longitudinal reconnection of pressure into tubes following the legs of the Λs. This is an example in which low pressure ceases to follow the coherent vortices.[3] Observe also that, because of self-induction, the arrows of the Λs get rounded and take on a hairpin shape, as do the boundary-layer vortices. For $M_c^{(i)} = 1$ (Figure 7.3), an analogous temporal evolution is observed, although it is strongly delayed and the level of turbulence generated is weaker. Vortex lines are much less convoluted than in the preceding case, and pressure structures are of smaller section. At the end of the simulation ($t = 140$), we have in fact the same structure as the one found by Sandham and Reynolds with their translative-type initial state. A side view of the vortex lines at this time

---

[3] This shows the difficulty of coherent-vortex recognition and confirms the need to have several visualization devices to understand turbulent flow dynamics.

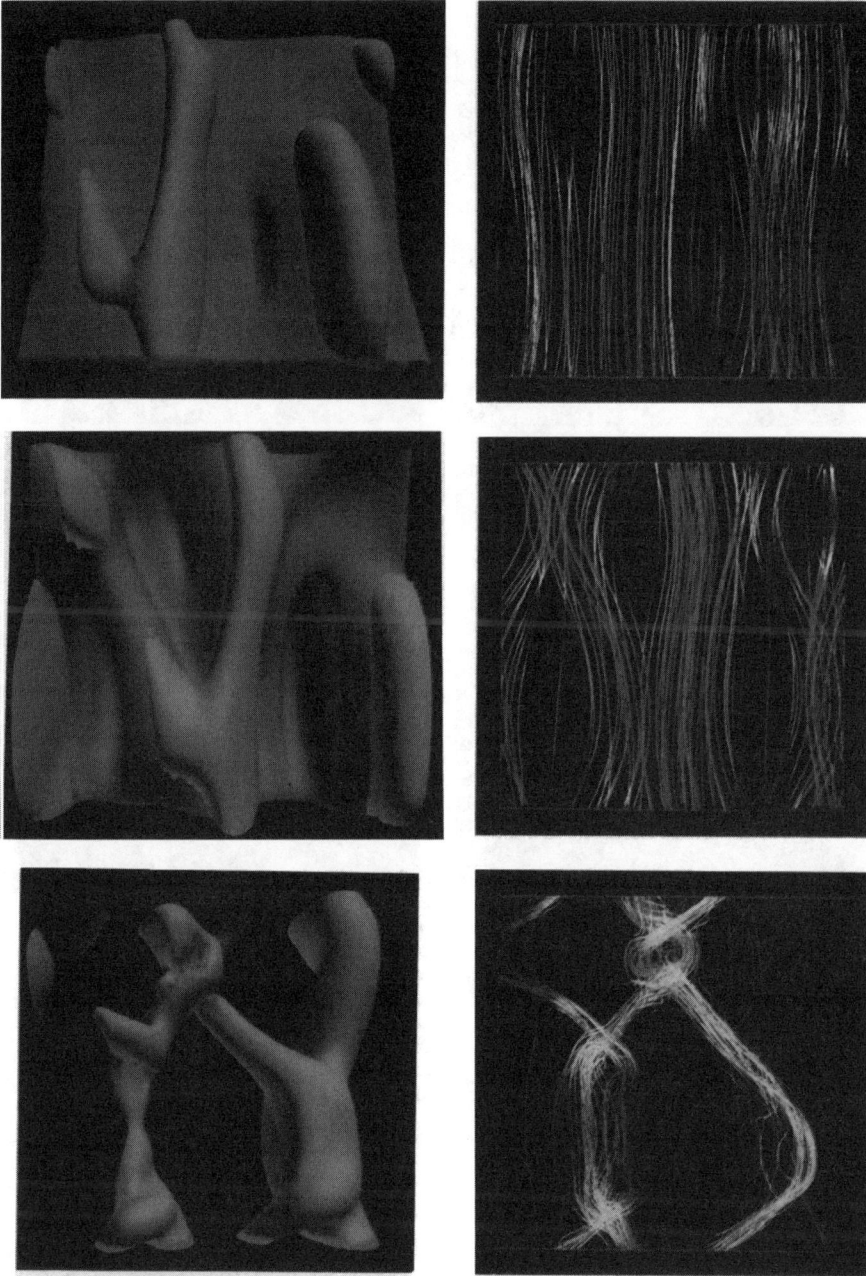

Figure 7.1. Top view of low-pressure isosurfaces (left) and vortex lines (right) in the DNS of a compressible temporal mixing layer at initial convective Mach number of 0.3 and at times (in units of δ_i/U, from top to bottom) 25, 45, and 50. (From Fouillet [98].)

(presented in Figure 7.4) shows that the two Λ vortices are stretched in parallel planes. They remain far from each other, in particular at the level of their tips, although the pressure structures are longitudinally reconnected. The absence of pairing confirms the quasi-total inhibition of KH instability

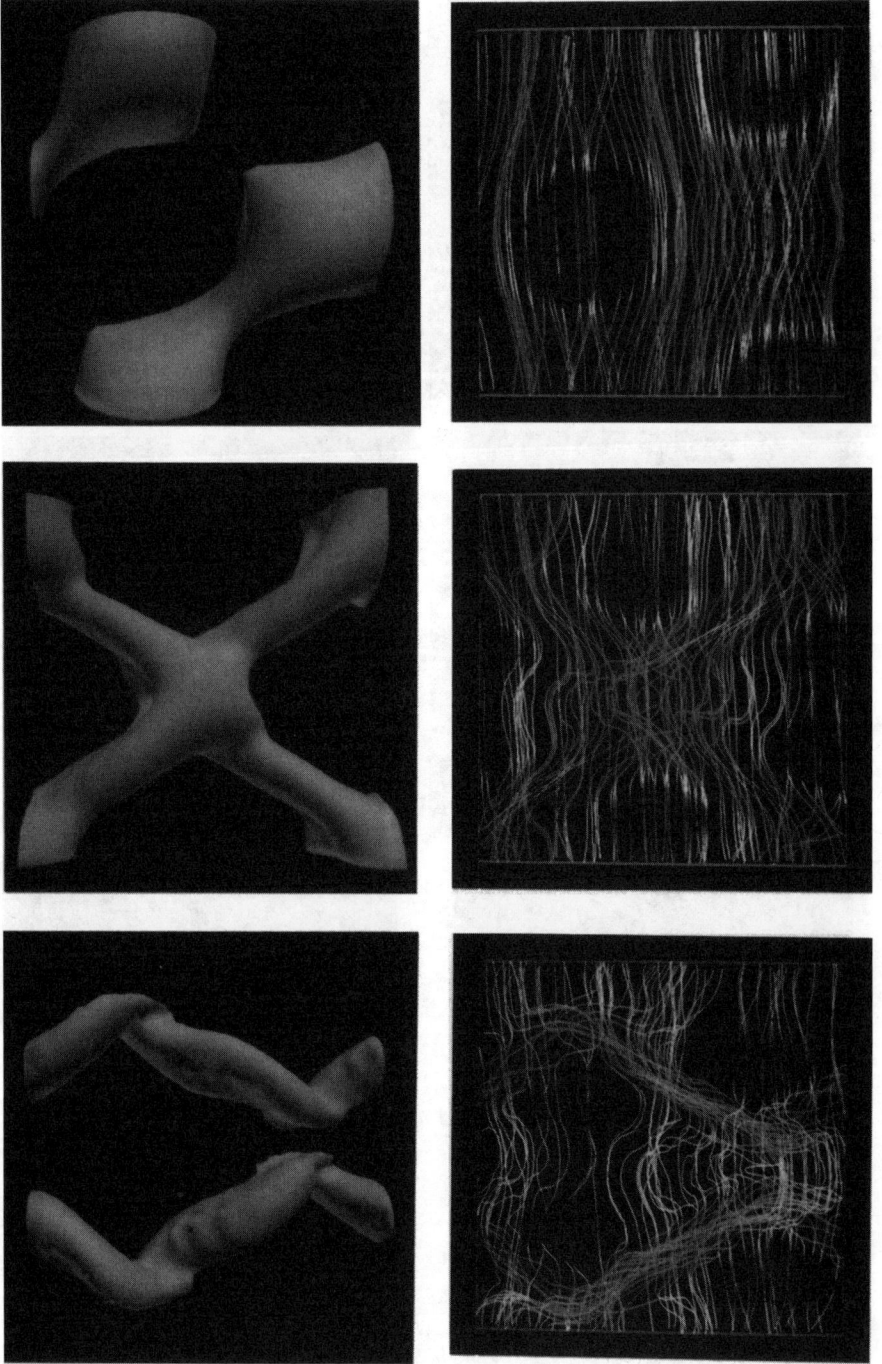

Figure 7.2. Same as Figure 7.1 at $M_c^{(i)} = 0.8$ for times (in units of δ_i/U, from top to bottom) 50, 60, and 100. (From Fouillet [98].)

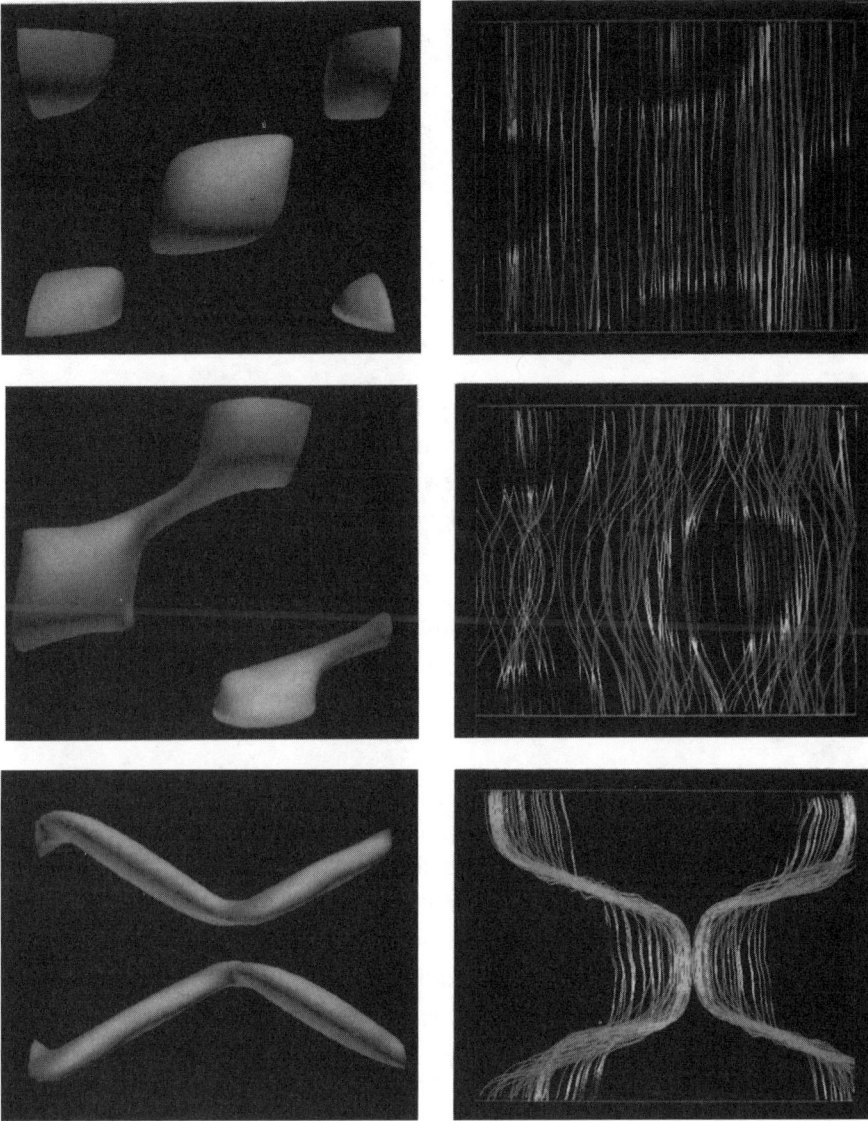

Figure 7.3. Same as Figure 7.1 at $M_c^{(i)} = 1$ for times (in units of δ_i/U, from top to bottom) 100, 120, and 140. (From Fouillet [98].)

by compressibility as soon as the convective Mach number approaches unity. This is certainly the reason why compressibility inhibits the spreading rates. By contrast, it favors the formation of longitudinal vortices. In fact, helical pairing in Fouillet's calculations is inhibited above $M_c^{(i)} = 0.6$–0.7.

A question arises about the time persistence of these coherent vortices. Pantano and Sarkar [229] have pointed out that (even in the weakly compressible case) the structures in a temporal-mixing layer DNS did not persist at very large times and eventually broke up into disordered turbulence not far from

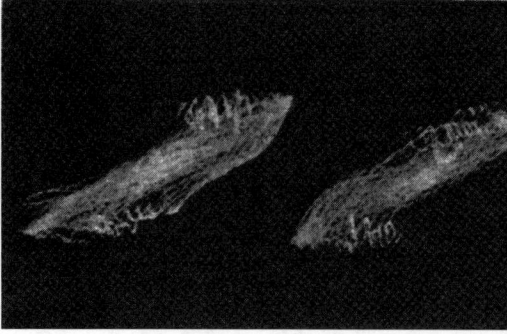

Figure 7.4. Side view (in the xy plane) of the vortex lines at $t = 140$ for $M_c^{(i)} = 1$. (From Fouillet [98].)

isotropy. The same was found in very well resolved LES initially involving twelve fundamental wavelengths carried out by Beer [20]. Figures 7.5 and 7.6 present for the latter simulations a view of the developed state for convective Mach numbers of, respectively, 0.2, 0.4 (Figure 7.5), 0.6, and 0.8 (Figure 7.6). It is clear that increasing the convective Mach number significantly reduces the size of turbulent structures.

However, box finite-size effects might contribute to the observed three-dimensionalization. It is therefore of interest to return to spatially growing mixing-layer simulations.

7.2.3 Spatial mixing layer

Comte et al. [53] also performed with the same code-underresolved DNS of three-dimensional, spatially growing, compressible mixing layers at a resolution of $80 \times 40 \times 30$. In the first simulation, the Mach numbers of the two currents are respectively 1 and 0.27, and the convective Mach number is 0.29. A weak, three-dimensional, random perturbation, imposed upstream on the constant average velocity $(U_1 + U_2)/2$ and the hyperbolic-tangent profile is regenerated at each time step. Figure 7.7 shows in this case the low-pressure isosurfaces (dark gray) with an isosurface of the average density $(\rho_1 + \rho_2)/2$ (light gray) materializing the interface between the two layers. This is obviously the same type of helical-pairing configuration we have obtained in this book with incompressible LES in wide domains, giving rise to branchings and local dislocations similar to the incompressible experimental findings of Browand and Troutt [32]. This gives us some confidence in the simulation of Figure 7.7 and validates in some sense the numerical method used. The simulation at $M_c = 0.7$ (Figure 7.8) has a totally different structure. One observes a pattern of elongated staggered Λ vortices very similar to what was obtained in the temporal case at $M_c^{(i)} = 0.8$.

More recent results come from the work of Doris [73], who studied the spatial mixing layer at $M_c = 0.64$ and $M_c = 1$ using a mixed model. The

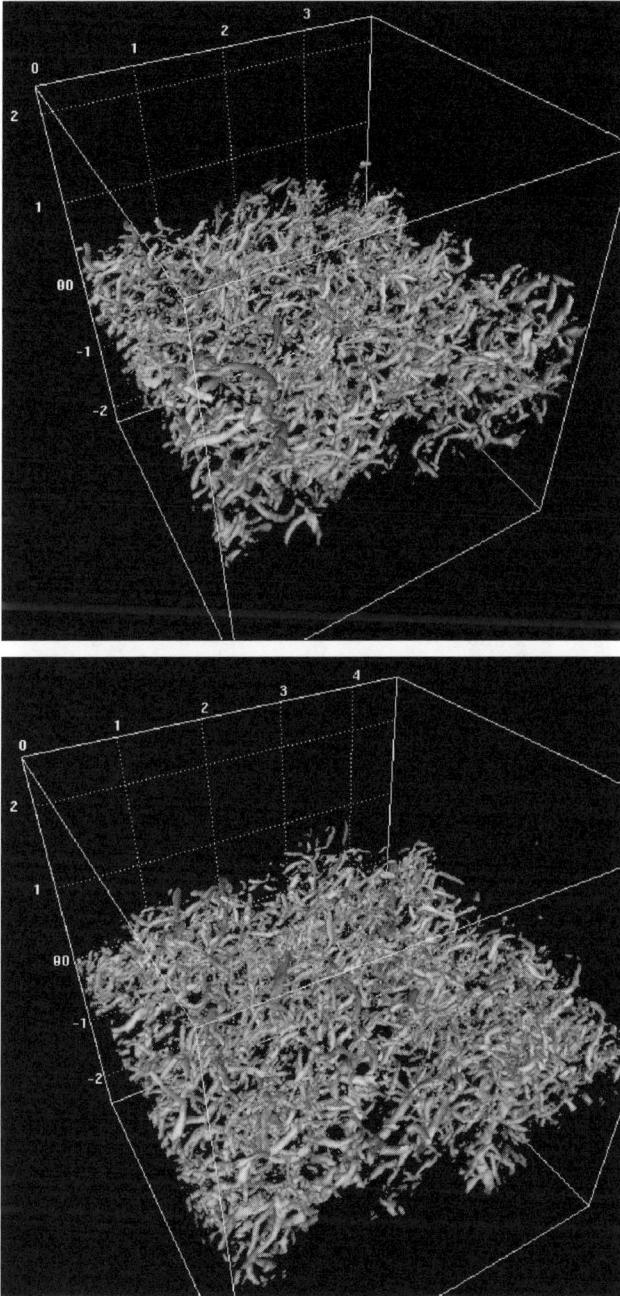

Figure 7.5. Perspective view of Q colored by longitudinal velocity in a developed, turbulent, temporal mixing layer LES carried out by Beer [20]. (Top) $M_c^{(i)} = 0.2$; (bottom) $M_c^{(i)} = 0.4$.

upstream random forcing is quasi-two-dimensional with a longitudinal velocity ten times larger than the spanwise one and no energy in the third direction. Vortices are visualized with the aid of the Q-criterion. At $M_c = 0.64$, helical pairing is observed with various amplitudes of the upstream forcing. At

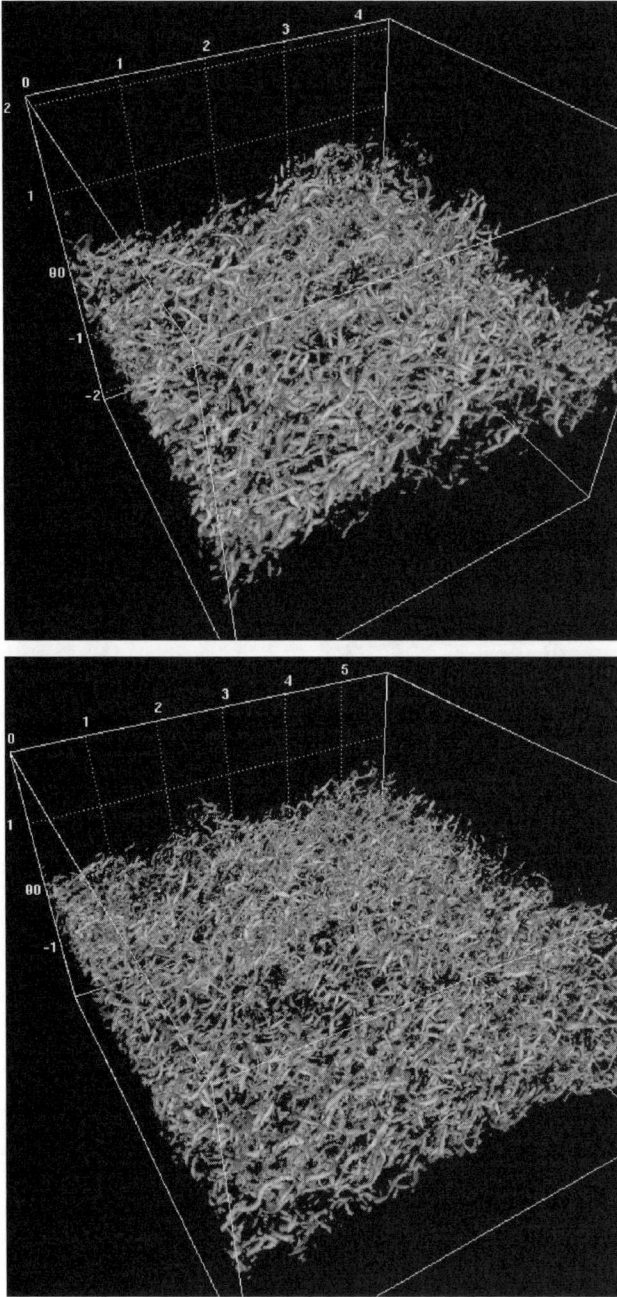

Figure 7.6. Same as Figure 7.5 for (top) $M_c^{(i)} = 0.6$ and (bottom) $M_c^{(i)} = 0.8$. (Courtesy A. Beer.)

$M_c = 1$, Doris notes the existence of large Λ-shaped structures without helical pairing within a highly complex three-dimensional flow. Various statistical quantities do not compare very well with the experiments of De Bisschop et al. [67] and Chambres [37].

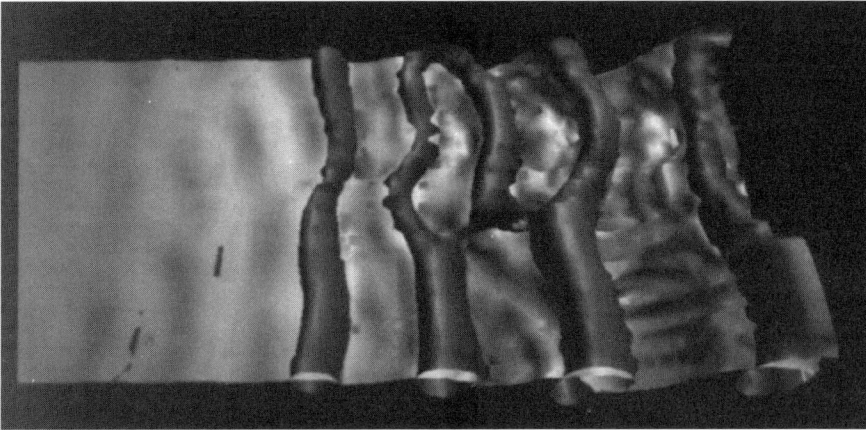

Figure 7.7. Weakly compressible spatial mixing layer at $M_c = 0.3$ showing isobaric low-pressure isosurface (dark gray) and isopycnal interface (light gray). (From Comte et al. [53].)

7.2.4 Compressible round jets

We show now several LESs of nonheated compressible round jets at Mach 0.7 and 1.4 carried out by Maidi and Lesieur in Grenoble. The jets are initiated by an upstream velocity profile, as in Chapter 6, and do not exit from real nozzles. Thus, in the supersonic regime, shocks and Mach waves, which are an important source of noise (screech noise in particular), cannot be obtained in these simulations. They are, however, interesting as far as coherent-vortex dynamics is concerned. Reynolds (Re) and Mach (M) numbers are based on quantities defined upstream (velocity at the jet center, jet diameter, and temperature). The subgrid-scale model is the FSF model. Boundary conditions of the Poinsot–Lele type [240] are used on the sides and downstream of the computational domain. To absorb downstream parasitic reflections, a sponge zone has been introduced downstream.

Code validation

The LES code was first validated by statistical comparisons with DNS of Freund [101] and the experiment of Stromberg et al. [276]. The upstream

Figure 7.8. Spatial mixing layer at $M_c = 0.7$. Shown are positive (white) and negative (black) vorticity components at a threshold of 10% of the upstream maximal vorticity with the isopycnal interface (light gray). (From Comte et al. [53].)

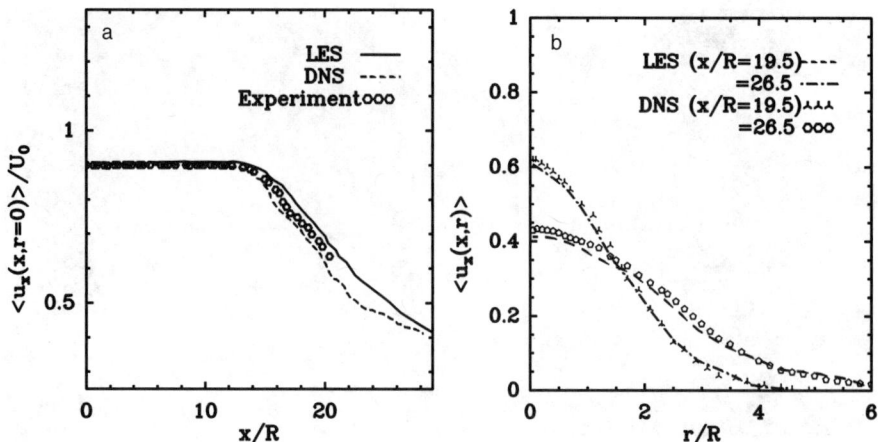

Figure 7.9. Forced compressible jet at Mach 0.9 and a Reynolds number of 3,600 of the LES with Freund's DNS and Stromberg et al.'s experiment. Mean longitudinal velocity is shown (a) at the jet centerline as a function of x and (b) as a function of r for various x.

velocity profile is forced here by the following periodic excitation:

$$U(r) = \frac{U_0}{2} \left[1 - \tanh \left(2.8 \left(\frac{r}{R} - \frac{R}{r} \right) \right) \right] \left[1 + \epsilon \sin \left(2\pi \frac{Str_D U_0}{D} t \right) \right],$$

(7.47)

with

$$Str_D = 0.45, \quad \epsilon = 0.0025, \quad M = 0.9, \quad Re = 3,600.$$

A nonuniform Cartesian grid of $100 \times 74 \times 74$ points is used. Figure 7.9(a) shows mean longitudinal velocity profiles on the jet centerline as a function of x/R obtained in our LES, in Freund's DNS [101], and in the experiment of Stromberg et al. [276]. Figure 7.9(b) displays mean longitudinal velocity as a funtion of r for various downstream distances x. The two simulations are very close to the experiment. This shows that LES is a good and cheap tool for simulations of low-Reynolds-number flows.

Subsonic free jet at $Re = 36,000$

The upstream velocity profile is still of the hyperbolic-tangent type:

$$U(r) = \frac{U_1 + U_2}{2} + \frac{U_1 - U_2}{2} \tanh \left[\left(\frac{R}{\theta} \right) \left(\frac{r}{R} - \frac{R}{r} \right) \right],$$

(7.48)

where U_1 is the velocity at the jet center and U_2 is a weak coflow of the same sign.[4] The two other velocity components are zero. This velocity profile is perturbed by a three-dimensional isotropic white noise. The Mach number is 0.7,

---

[4] Coflows are used in turbojet engines as a way to reduce noise.

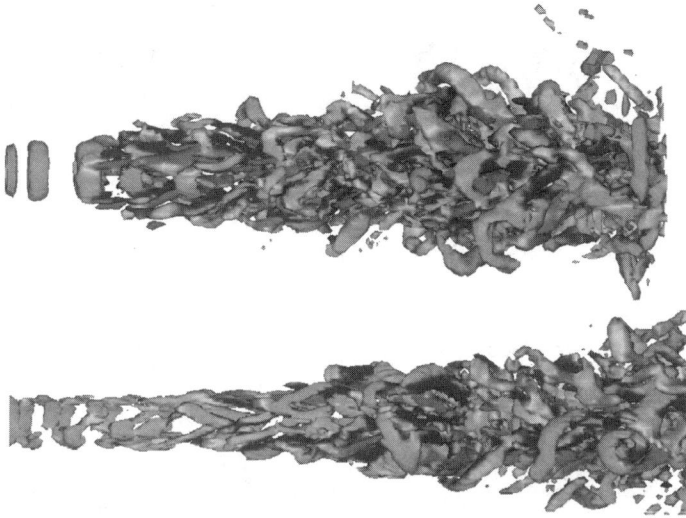

Figure 7.10. (Top) Free jet at Mach 0.7 and a Reynolds number of 36,000. The Q isosurfaces are colored by longitudinal vorticity. (Bottom) The same jet at Mach 1.4. (Courtesy M. Maidi.)

and so the convective Mach number is ≈ 0.35. This is weak in comparison with values of the order of 0.6–0.7 at which compressibility effects start being important. Because the jet is nonheated, one may expect a jet behavior close to constant-density jets. Figure 7.10 (top) shows at a time of $300D/(U_1 - U_2)$ positive-Q isosurfaces colored by longitudinal vorticity. One sees clearly upstream the shedding of quasi-axisymmetric vortex rings stretching longitudinal vortices just before the appearance of alternate pairing. Hence the three-dimensional perturbation caused by fluid lateral ejection owing to longitudinal vortices seems to trigger alternate pairing of vortex rings. More downstream, the flow contains a very chaotic superposition of alternate pairing of rings and small-scale–developed turbulence. From a statistical point of view, the flow in the jet central region is close to isotropic turbulence.

This LES can be validated against experimental measurements of Hussein et al. [131] obtained for the jet self-similar region. Figures 7.11(a)–7.11(d) display radial profiles of, respectively, the mean longitudinal velocity and Reynolds stresses at four different downstream distances between $x/R = 27$ and $x/R = 28.5$. One sees a good merging of the profiles, which confirms similitude within this fully developed turbulence region.

Supersonic free jet at $Re = 36,000$

For a Mach number of 1.4, which corresponds to a convective Mach number of ≈ 0.7, Figure 7.10 (bottom) presents the jet dynamics at $t = 300D/(U_1 - U_2)$ visualized by Q and longitudinal vorticity. We see the quasi-total disappearance of the axisymmetric mode, and vortex rings exhibit helical pairing immediately downstream of the nozzle. This is accompanied by an intense reduction

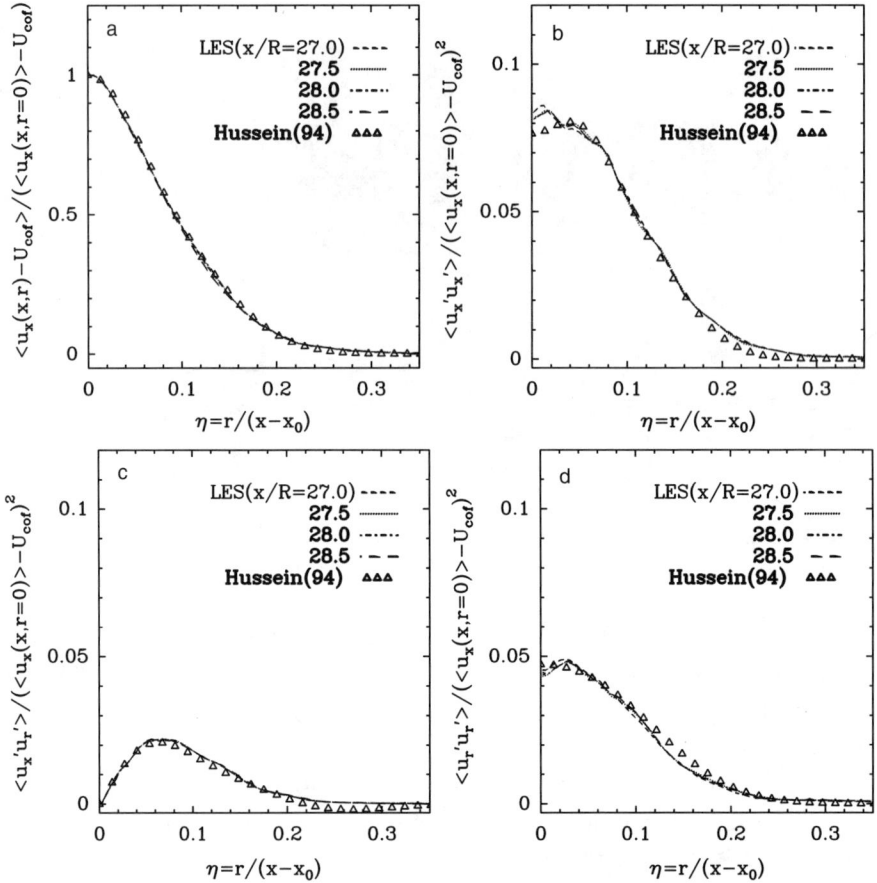

Figure 7.11. Same jet as in Figure 7.10 for developed region. Comparisons of LES with Hussein et al.'s experiments: (a) longitudinal velocity; (b) longitudinal Reynolds stress; (c) shear stress; (d) radial Reynolds stress. (Courtesy M. Maidi.)

of the jet spreading rate, in agreement with three-dimensional linear-stability analyses of the mixing layer carried out by Sandham and Reynolds [253]. Further downstream, the jet suddenly three-dimensionalizes in the small scales and starts increasing at rates close to the subsonic regime. However, at any given downstream distance, the jet is much more confined than in the subsonic case. Recall that DNS of three-dimensional plane mixing layers show an inhibition of vortex pairing above a convective Mach number of the order of 0.6 to 0.7. No such effect is observed for these axisymmetric mixing layers in terms of alternate pairing, and higher Mach number LESs are needed.

We show now some statistics in this case, which have been taken over a period of $300D/(U_1 - U_2)$, which is sufficient to have statistical convergence. Figure 7.12(a) presents a comparison for the two Mach numbers 0.7 and 1.4 of the mean longitudinal velocity on the axis as a function of x/D. The velocity drop around $x/D = 5.5$ at Mach 0.7 indicates the end of the potential core.

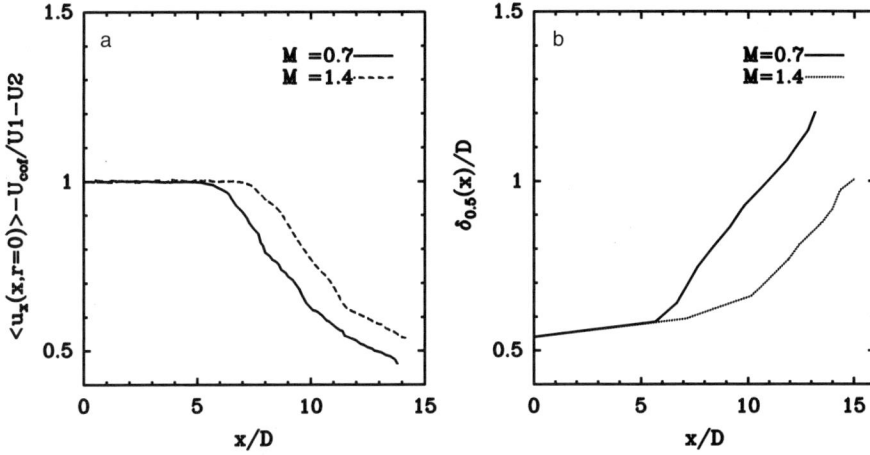

Figure 7.12. Comparisons between the subsonic and supersonic jets for (a) longitudinal velocity and (b) jet thickness.

The potential-core length is thus $5.5D$ in this case. It is $7.7D$ at Mach 1.4, which is an increase of 27%. The spatial evolution with x of the jet width is presented on Figure 7.12(b). We see that both simulations give the same result up to $x/D \approx 5.5$. Then the subsonic jet transitions to developed turbulence and spreads linearly more rapidly. The supersonic jet widens at a weaker rate up to $x/D \approx 10$. Afterward it develops into turbulence, with the same growth rate as the subsonic jet. Comparison of the Reynolds-stress profiles for the two Mach numbers are given in Figure 7.13. These quantities have been calculated at the end of the potential core. Shear ($\langle u_x' u_r' \rangle$), radial ($\langle u_r' u_r' \rangle$), and azimuthal ($\langle u_\theta' u_\theta' \rangle$) components are strongly reduced by compressibility. In contrast, the axial component $\langle u_x' u_x' \rangle$ is weakly affected and even increases

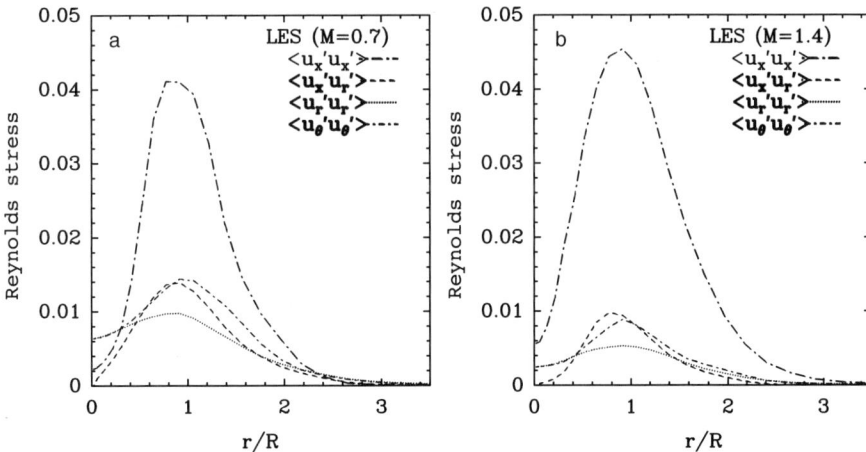

Figure 7.13. Radial profiles of Reynolds stresses at the end of the potential core at (a) Mach 0.7 and (b) Mach 1.4.

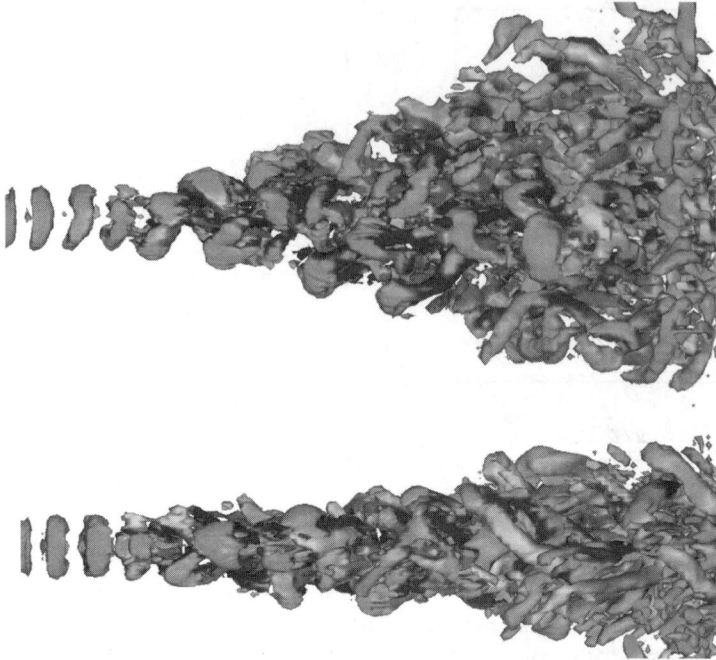

Figure 7.14. Harmonic–subharmonic varicose-flapping forcing at Mach 0.7 and a Reynolds number of 36,000. Isosurfaces of Q are colored by longitudinal vorticity. (Top) View of the bifurcating plane. (Bottom) View of the bisecting plane.

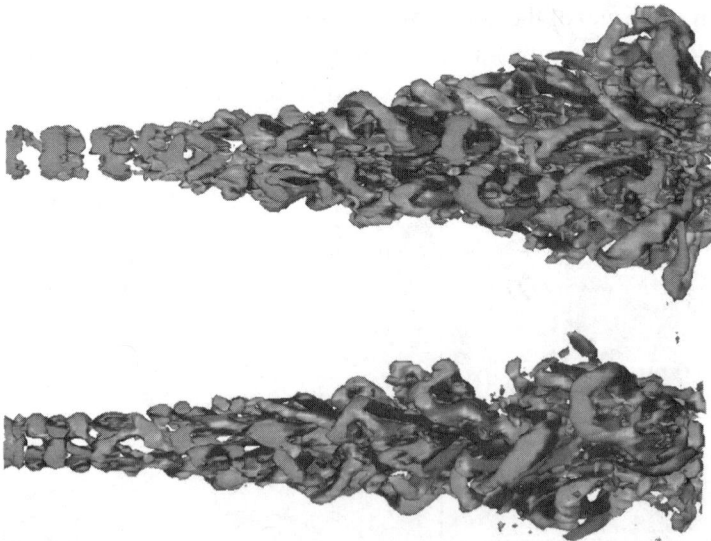

Figure 7.15. Harmonic–subharmonic varicose-flapping forcing at Mach 1.4 and a Reynolds number of 36,000. Isosurfaces of Q are colored by longitudinal vorticity. (Top) View of the bifurcating plane. (Bottom) View of the bisecting plane.

slightly. This agrees with the experiments of Goebel and Dutton [112] and Urban and Mungal [283] and with the DNS of Freund et al. [102]. However, the term $\langle u'_x u'_x \rangle$ decreases when the Mach number increases in the experiments of Samimy and Elliott [85, 252].

Varicose-flapping jet control

Here we show effects on the jet of a varicose-flapping excitation such as considered in Chapter 6, where the varicose mode is harmonic and the flapping mode is subharmonic.

At Mach 0.7, we take $Str_D = 0.39$, which corresponds to the frequency of vortices at the end of the potential core for the free subsonic jet. The Reynolds number is $Re = 36,000$. Figure 7.14 displays in this case Q isosurfaces colored by longitudinal vorticity in the bifurcating plane and in the bisecting plane.

At Mach 1.4, we have $Str_D = 0.44$. Figure 7.15 shows the resulting flow. The delay in the development of KH instability is recovered.

7.3 Weakly compressible wall flows

7.3.1 LES of spatially developing boundary layers

Lesieur ([170], pp. 450–452) reported a LES by Ducros et al. [80] of a temporal (periodic in the flow direction) compressible boundary layer on an adiabatic wall at a Mach number at infinity of $M_\infty = 4.5$. The simulation was done using the standard SF model, which enables in this case the simulation to continue beyond transition. However, as already stressed, this model does not work for transition in a boundary layer at low Mach number (or for incompressible flow) at which, like the Smagorinsky model, it is too dissipative and prevents small perturbations from degenerating into turbulence. Conversely, this model has been used with success in its filtered version (FSF model) for the simulation of $M_\infty = 0.5$ boundary layer of an ideal gas developing spatially over an adiabatic flat plate with a low level of upstream forcing (Ducros et al. [81]). We present this work here, in which compressibility effects are very weak close to the wall, as a way to better understand the very difficult problems of transition to turbulence and the nature of developed turbulence in an incompressible boundary layer without a pressure gradient. In the work of Ducros et al. [81], periodicity is assumed in the spanwise direction. Non-reflective boundary conditions (based on the Thompson characteristic method [279]) are prescribed at the outlet and the upper boundaries. With the minimal resolution of $650 \times 32 \times 20$ resolution points in the streamwise (x_1), transverse (x_2), and spanwise directions (x_3), respectively, covering a range of

streamwise Reynolds numbers $Re_x \in [3.4 \times 10^5, 1.1 \times 10^6]$, transition has
been obtained for 80 hours of time processing on a CRAY2 machine (whereas
DNS of the entire transition takes about ten times longer). The flow upstream
is the superposition of the laminar profile at this Mach, a two-dimensional
perturbation forcing the most amplified Tollmien–Schlichting mode, and a
three-dimensional white noise such that

$$U(0, x_2, x_3, t) = U_{\text{lam}}(x_2) + 5 \times 10^{-3} \hat{U}(x_2) + 8 \times 10^{-3} U_{\text{rand}}(x_2, x_3, t),$$
$$(7.49)$$

where $U_{\text{lam}}(x_2)$ is the laminar profile of the similarity equations, $\hat{U}(x_2)$ is the
most amplified eigenmode of the two-dimensional Tollmien–Schlichting (TS)
waves, and $U_{\text{rand}}(x_2, x_3, t)$ is a randomly chosen, three-dimensional, white
noise of variance U_∞^2. The upstream Reynolds number based on the displacement thickness is $R_{\delta_1} = 1,000$. This is supercritical with respect to the
critical value of 520 predicted by the linear-stability theory in the incompressible case, which justifies the TS wave contribution in the upstream field. It
should be stressed that simulating a complete boundary-layer[5] transition (up
to developed turbulence), starting upstream from a subcritical laminar profile perturbed by a weak three-dimensional perturbation, is not possible right
now because of the excessive computational cost in simulating the very slow
growth of the viscous instabilities involved.

We return to the Ducros et al. [81] LES, where we see how the TS wave
generated upstream propagates downstream. First, quasi-two-dimensional billows of relatively low pressure and high vorticity form and travel with the wave
velocity. A top view of the low-pressure and longitudinal vorticity in the transitional region is shown in Figure XII-9 in Lesieur ([170], p. 400). During the
transition, these rolls evolve into a staggered pattern that breaks down into
turbulence. Meanwhile, the longitudinal velocity develops weak streaks close
to the wall, as shown in Figure XII-10 in Lesieur ([170], p. 401). These streaks
are the seed of stronger low- and high-speed streaks in the developed region.
We reproduce in Figure 7.16 (taken from Ducros et al. [81]; see also Lesieur
[170]) a view of a very strong hairpin ejected away from the wall just at the
beginning of the developed turbulent region. The numbers on the x-axis correspond to the downstream distance in units of δ_i, the upstream displacement
thickness. Downstream, there are several series of hairpins ejected above the
low-speed streaks. However, the resolution close to the wall is insufficient
(first point at $y^+ = 5$–6) to allow for good predictions of average quantities
such as the friction coefficient at the wall or the shape factor. An interesting
feature of the hairpin shown in the figure is the following asymmetry: Vortex
lines are much more condensed in the right leg than in the left one, and, as

---

[5] Here we refer to a boundary layer without a longitudinal pressure gradient.

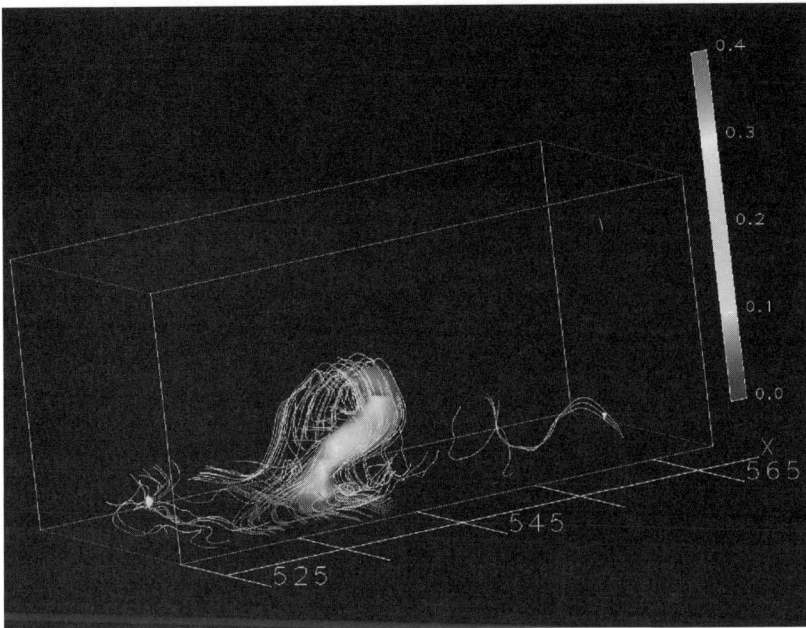

Figure 7.16. LES of the spatial boundary layer at Mach 0.5 showing vortex lines and low pressure characterizing a hairpin vortex ejected from the wall at the beginning of the developed region.

a consequence, only the right leg induces a marked pressure trough. As far as the vorticity is concerned, this right leg forms in fact a quasi-longitudinal vortex close to the wall that rises above owing to self-induction, forming a semiarch. Similar traveling quasi-longitudinal vortices will be visualized in animations of a channel later in the chapter (see section 7.3.2).

Although it gives interesting qualitative information on the structure of turbulent boundary layers, the LES just described does not have a sufficient resolution close to the wall, as just mentioned. Here, we present new results with a finer resolution at the wall ($y^+ = 1$ or 2), at a lower Mach number at infinity (0.3). It is known that transition in the boundary layer on a flat plate depends on the type of perturbations exerted upstream on the flow (see Lesieur [170]). In Klebanoff et al. [143], the boundary layer was forced upstream with a thin metal ribbon parallel to the wall and stretched in the spanwise direction, which vibrates two dimensionally close to the wall. In this experiment, the three-dimensional forcing (obtained with the aid of tape fixed on the wall and regularly spaced in the spanwise direction) was harmonic. This corresponds to what is referred to as the K-mode, where the crests of the TS waves oscillate in phase in the spanwise direction. In contrast, if the perturbation is subharmonic, the crests oscillate out of phase. This is called the H-mode, from Herbert ([121]; see also Kachanov and Levchenko [139]), and it corresponds to a staggered organization of vortex filaments. Herbert could

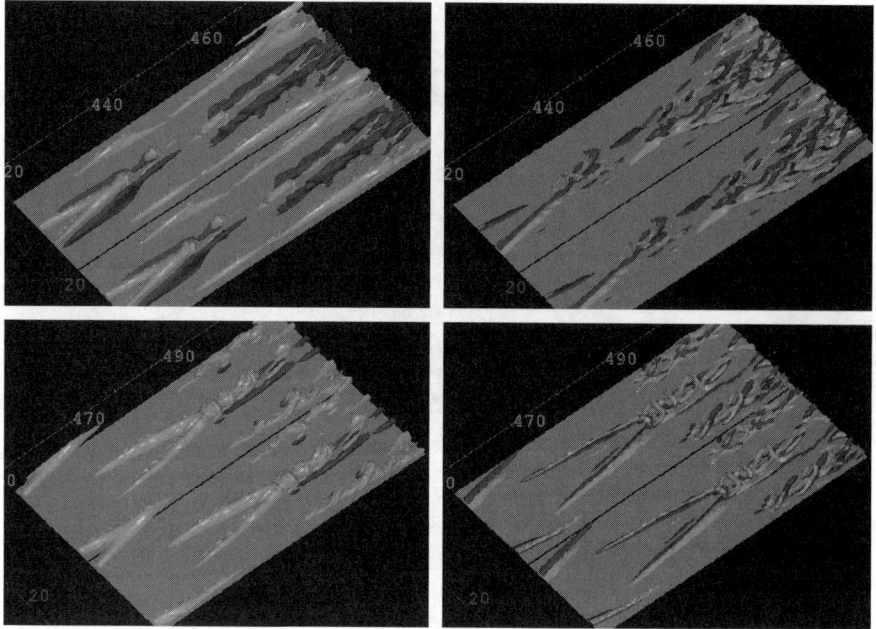

Figure 7.17. LES of a spatial boundary layer at Mach 0.3; top and bottom panels show K- and H-transition respectively; the l.h.s. and r.h.s. correspond, respectively, to velocity and vorticity fluctuation components (dark, positive; light gray, negative); dark gray marks isosurfaces of positive Q. (Courtesy E. Briand.)

show for the temporal problem[6] that the staggered mode was more amplified than the aligned mode. In fact, Ducros et al. [81] observed a transition of the subharmonic type.

We present now the LES of the Mach 0.3 spatially developing boundary layer over a flat plate carried out by Briand [31]. It is started here with a different set of nonlinear parabolized stability expansion (PSE) calculations from Bertolotti and Herbert [24] and Herbert [122] (see also Airiau [5]). To this upstream state (with $Re_{\delta_1} = 1,000$), one superposes a three-dimensional white noise of amplitude 0.2, the amplitude of the PSE perturbation. The following discussion is usefully complemented by Animations 7-1 and 7-2 on the CD-ROM. In the K-mode case, one sees formation in the transitional region of large, longitudinal, Λ-shaped vortices lying on the wall that are in phase in the flow direction (see Figure 7.17, top). In the H-mode case, the vortices are staggered (see Figure 7.17, bottom). The figures show at the end of transition the longitudinal components of velocity and vorticity and also positive Q. One sees that the Λ vortices are very well correlated with a system of induced high- and low-speed streaks.[7] Notice also on the vorticity plots

[6] In Herbert's analysis a secondary-instability analysis is used where a three-dimensional perturbation is superposed on a TS wave of finite amplitude.
[7] This is not apparent on the figure for the H-mode case owing to an ill-chosen threshold.

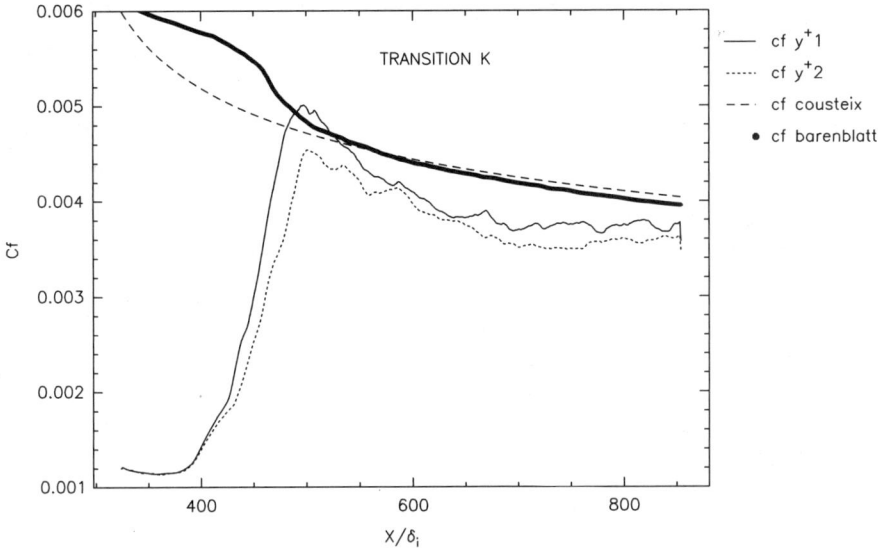

Figure 7.18. LES of a spatial boundary layer at Mach 0.3. Friction coefficients are plotted against downstream distance and are compared with theoretical predictions of Van Driest and Barenblatt and Protokishin. (Courtesy E. Briand.)

that the big Λs induce "antivorticity" close to the wall owing to the zero velocity condition at the wall.[8] Downstream of $\approx 440\delta_i$, the streaks become purely longitudinal. This is accompanied by the fast shedding of small arch vortices ejected from the tip of the Λs, as indicated by Q isosurfaces.

Figure 7.18 shows for the K-transition the downstream evolution of the friction coefficient at the wall in comparison with the theoretical predictions of Van Driest (discussed in Cousteix [56]) and Barenblatt and Prostokishin [13]. One sees a good agreement of the LES with these predictions with a resolution of $y^+ = 1$ improving the result. It is even better in the H-mode case. The peak in the friction coefficient is at $490\delta_i$, which is much farther downstream than the change of regime of the velocity streaks, and might be associated with an event such as the localized creation of a big hairpin vortex observed in the simulations of Ducros et al. [81] and presented in Figure 7.16. Figure 7.19 shows for the K-mode case (but results are very similar in the H-mode case) the rms longitudinal velocity component u' at a downstream distance such that $Re_{\delta_1} = 1,670$, compared with Spalart's [273] DNS at Reynolds numbers of 1,000 and 2,000. Again, the agreement is good, since the results lie between Spalart's predictions. Animation 7-1 concerns both transition and developed turbulence with the K-forcing. It shows first isosurfaces of u', ω'_x, and ω'_z

---

[8] Indeed, a vortex approaching a wall will create under it a velocity gradient at the wall of vorticity having a sign opposite to the vortex's vorticity. But this vorticity may be too weak to induce rotation of the fluid around it, in such a way that antivorticity is not always organized into a coherent vortex.

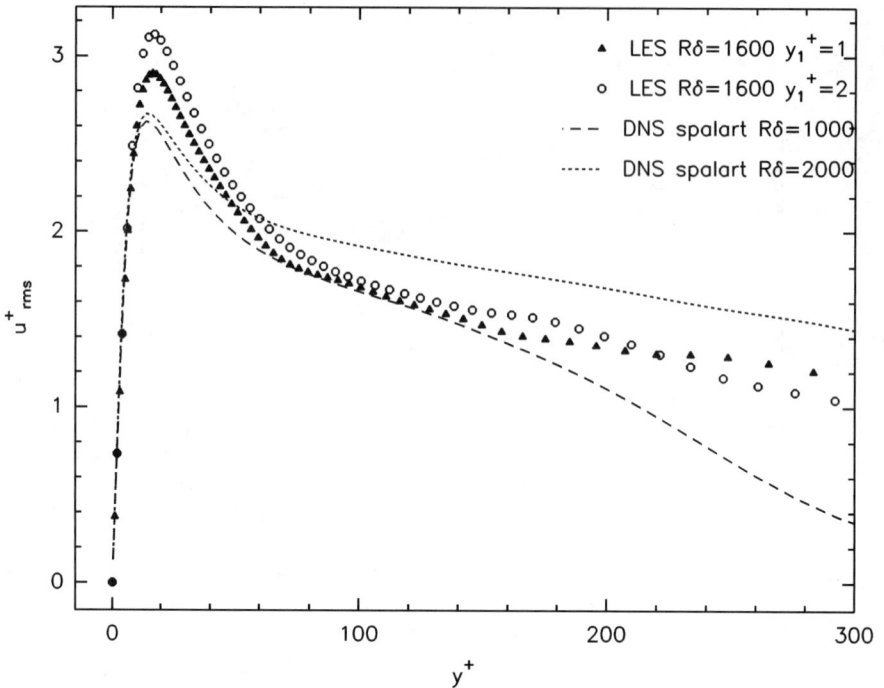

Figure 7.19. Spatial boundary layer at Mach 0.3 comparing the rms velocity fluctuations with Spalart's DNS. (Courtesy E. Briand.)

during transition and confirms the passage of the powerful Λ vortices lying on the wall. As already stressed, these induce high- and low-speed velocity streaks and positive antivorticity. They also induce regions of high spanwise vorticity ω_z (and hence high friction), which should be correlated with high-speed streaks. Later, in the region of developed turbulence, the movie shows the travel of velocity streaks and then of low-pressure isosurfaces. Above the low-speed streaks, structures are ejected, and these might well be the footprint of ejected hairpins. The low-pressure animation shows only the head of these hairpins. In fact, there are several hairpins above a single low-speed streak, as was already observed in Ducros et al. [81]. In this respect, we no longer have the perfect correlation between hairpins and streaks that we observed during the transitional stage. It is therefore difficult in the developed region to associate the streaks to a system of purely longitudinal vortices at the wall.

Animation 7-2 shows isosurfaces of u' and Q for K- and H-forcings. For K-transition, two values of the Q threshold are compared (0.01 and 0.02) close to the tip of the big Λ vortex. It is interesting that the larger value allows us to display along the Λ legs a new hierarchy of arch vortices that are smaller in size than those ejected from the Λ tip. Then the movie presents u' and Q in the developed region from a frame moving with a velocity of $0.6U_\infty$. The various structures are approximately stationary, which tends to indicate that

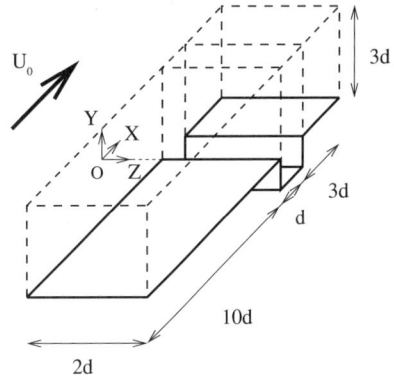

Figure 7.20. Sketch of the computational domain. (Courtesy Y. Dubief.)

they travel at a velocity of this order in the fixed frame. The rest of the movie concerns models of wave packets and will not be commented upon here (see Briand [31] for details).

7.3.2 Boundary layer on spanwise cavities

Deep cavity

We first look at the effect of a spanwise cavity (whose dimensions are typically of the order of the boundary layer thickness) on the vortical structure of a turbulent boundary layer. Such a configuration has recently generated renewed interest in the field of turbulence control (Choi and Fujisawa [39], Pearson et al. [232]). The cavity belongs to the category of passive devices able to manipulate skin friction in turbulent boundary layer flow. Depending on its dimensions, the drag downstream of the cavity can be increased or decreased.

To investigate the effects of a spanwise cavity on the near-wall structure of turbulent boundary layer flows, Dubief and Comte [76] (see also Lesieur et al. [172]) have performed a spatial numerical simulation of the flow over a flat plate with a spanwise square cavity embbeded in it. The goal here is to show the ability of LES to handle more complex geometries. The numerical code used is an evolution of the compressible Navier–Stokes solver COMPRESS, already discussed. The new version of the code, WOMBAT, written by Dubief [78], is multidomain, ensuring also spatial fourth-order accuracy at the domain borders. The subgrid model used is the FSF model but now in a four-point formulation in planes parallel to the wall. Periodicity is assumed in the spanwise direction. The computational domain, sketched in Figure 7.20, is here decomposed into three blocks. The Mach number at infinity is 0.5. The large dimension of the upstream domain is required for a proper generation of the inlet condition. The coordinate origin is located at the upstream edge of the cavity. The inlet, cavity, and downstream flat plate blocks have resolutions of, respectively, $101 \times 51 \times 40$, $41 \times 101 \times 40$, and $121 \times 51 \times 40$.

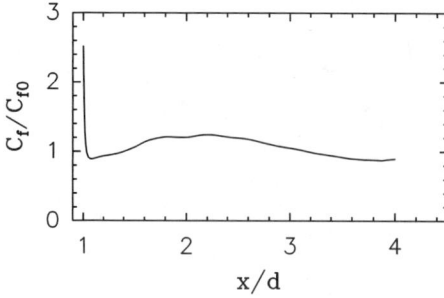

Figure 7.21. Mach 0.5 cavity showing the longitudinal evolution of the skin friction coefficient normalized by its smooth wall value. (Courtesy Y. Dubief.)

The minimal grid spacing at the wall in the vertical direction corresponds to $\Delta y^+ = 1$ in terms of inlet boundary layer units. The streamwise grid spacing goes from $\Delta x^+ = 3.2$ near the cavity edges to 20 at the outlet. The spanwise resolution is $\Delta z^+ = 16$. The Reynolds number of the flow (based on $d = \delta$, the boundary layer thickness) is 5,100, which is similar to the intermediate simulation of Spalart [273] at $R_\theta = 670$. The inflow is generated using the aforementioned method of Lund et al. [192], which we will describe in more detail. This method is based on the similarity properties of canonical turbulent boundary layers. At each time step, the mean and fluctuating velocities, temperatures, and pressures are extracted from a plane, called the recycling plane, and renormalized at an appropriate inlet scaling. The method is based on the inner and outer scaling laws, where the variables scale with (v_*, ν) and (v_*, δ), respectively. The statistics found are in good agreement with Spalart's data. Figure 7.21 shows the distribution of the skin-friction coefficient C_f normalized by its smooth wall value on the downstream flat plate. Immediately downstream of the cavity the skin friction coefficient experiences a sharp rise, followed by a small undershoot below the upstream value. Then it rises again. It eventually relaxes toward its smooth wall value in an oscillatory manner. This behavior is consistent with previous experimental results of Pearson et al. [232] with a smaller d/δ ratio. The local drag reduction observed in the present simulations is smaller than that obtained by these authors, probably because of a larger size of the cavity in our case. The large magnitude of the skin friction at the edge is obviously caused by the hairpin-type vortices, generated in the upstream boundary layer, that travel above the cavity and impinge on its downstream ridge, as coherent-vortex analysis presented in the following will show. Spanwise correlations of u, v, and w (not shown here) indicate a slight change in the streak spanwise wavelength. In the buffer layer, whereas negative u correlation peaks at $z^+ = 50$ in the upstream boundary layer (giving the right streak spacing $\lambda_z^+ = 100$), the spanwise wavelength is reduced downstream of the cavity to $\lambda_z^+ = 70$. Figure 7.22 shows an instantaneous visualization of u fluctation isosurfaces. The vertical extent of low-speed streaks is increased as they pass over the cavity. The vorticity field is plotted using isosurfaces of

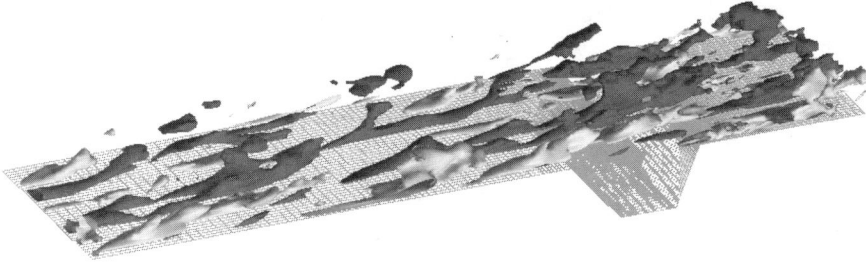

Figure 7.22. Mach 0.5 cavity showing isosurfaces of streamwise velocity fluctuations. Black, $u' = -0.17U_0$; white, $u' = +0.17U_0$. (Courtesy Y. Dubief.)

the vorticity norm conditioned by positive Q (Figure 7.23). Upstream, the quasi-longitudinal vortices creep up the wall and rise just like those present in the boundary layer on the flat plate studied earlier. Their length is of the order of 300 wall units, as in the experiments in turbulent boundary layers [8]. The structures downstream of the cavity are smaller and less elongated in the streamwise direction. It was checked that the statistics here show some sort of return to isotropy (see Dubief [77]) in terms of the famous Lumley map in the plane of components III and $-II$, the third and second invariants of the anisotropy tensor associated with the Reynolds-stress tensor (Lumley [190], Lumley and Newman [191]). Figure 7.24 gives a view of the flow within the cavity (where the upstream vertical wall is removed). Vortices aligned in the y direction can be isolated in the upstream part of the cavity. The curvature of the core of these vortices corresponds to the local curvature of the recirculating flow, and they may be due to Görtler instability. The flow inside the cavity is highly unsteady, and there is obviously a high level of communication between the recirculating vortex and the turbulent boundary layer.

We show now the LES of a flow at Mach 0.1 in a boundary layer on a flat plate passing over a square cavity deeper than the former one (Lesieur et al. [176]). Here, compressibility effects are negligible. The spanwise extent of the cavity is still $2d$, and periodicity is again assumed in the spanwise direction.

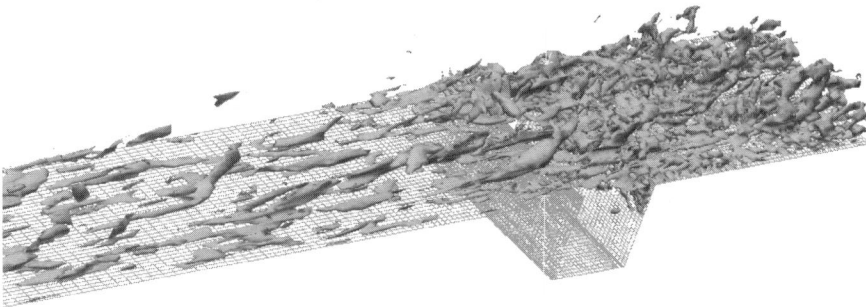

Figure 7.23. Mach 0.5 cavity showing isosurfaces of the vorticity norm filtered by positive Q. (Courtesy Y. Dubief.)

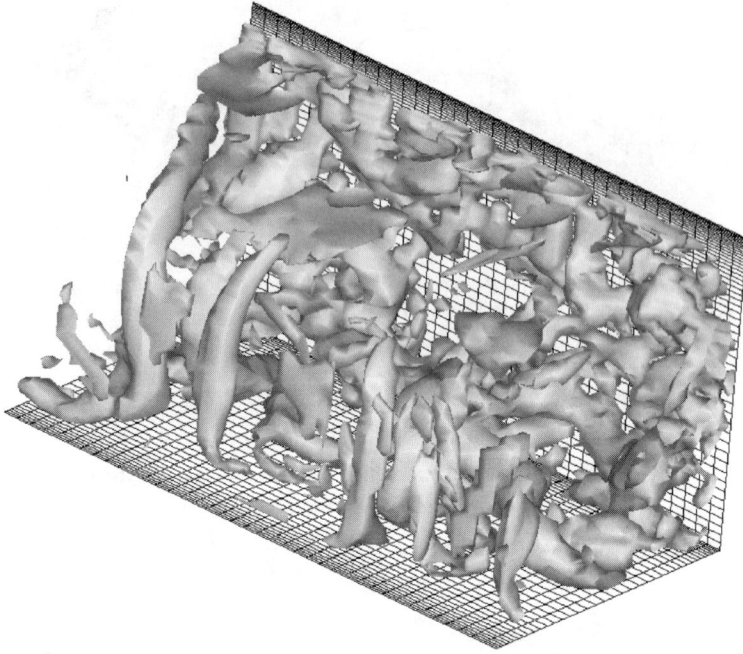

Figure 7.24. Mach 0.5 cavity showing vorticity norm filtered by positive Q inside the cavity. (Courtesy Y. Dubief.)

The Reynolds number is based on the velocity at infinity U_0 and the cavity depth d is 270,000. The total number of grid points is 688,000. The grid is refined close to the wall and in the region of strong shear between the edges. The upstream condition is less sophisticated than for the cavity just presented because it consists of the mean velocity of a turbulent boundary layer on a flat plate given by a power law of the form $u/U_0 = (y/\delta)^{1/7}$ proposed by Schlichting [257] with $\delta = 0.4d$. The total stress τ_0 is defined by Blasius's empirical relation

$$\frac{\tau_0}{\rho U_0^2} = 0.0225 \left(\frac{\nu}{U_0 \delta} \right)^{\frac{1}{4}}, \tag{7.50}$$

allowing us to define the friction velocity. A white noise of 5% intensity is superposed on this profile close to the wall. Figure 7.25 (bottom left) presents a vertical section of the vorticity norm. The pressure signal on a line $y = 0$, (i.e., the upstream and downstream walls and the line joining the two edges of the cavity) is shown in Figure 7.25 (top left). A three-dimensional plot of Q is presented in Figure 7.25 (top right) with a zoom in Figure 7.25 (bottom right). These figures indicate the passage of KH-type vortices shed behind the first backstep and impinging on the second edge of the cavity. The pressure signal at the wall has a high frequency associated with these vortices and a much lower one that might be due to the recirculation of the flow within the cavity. The vortices shed behind the first edge are quasi-two-dimensional, and, because

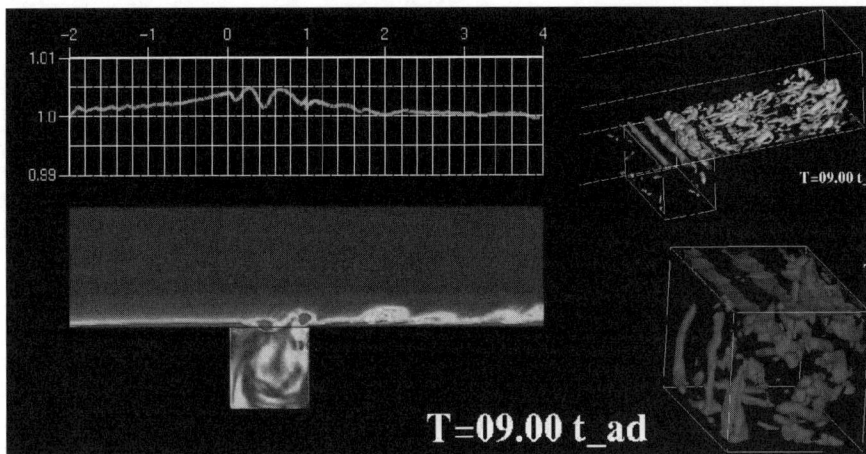

Figure 7.25. Mach 0.1 cavity. (Top left) Pressure at the wall; (bottom left) vertical section of vorticity norm; (top right) three-dimensional map of Q; (bottom right) zoom of Q in the cavity.

the length to travel is short, they do not have time to three-dimensionalize significantly. Then they impinge on the second edge and transform into large, very coherent Λ vortices, which become arches by raising their tips via self-induction. When looking at Q plots inside the cavity, we see on the top KH vortices passing by, whereas the recirculation in the cavity produces numerous longitudinal vortices, as in the former cavity presented. These longitudinal structures have also been observed experimentally in [198]. In fact the two cavities presented at Mach 0.5 and Mach 0.1 are very similar in structure within the cavity and downstream. Such numerical and experimental fundamental studies are extremely important for the automobile industry in terms of aeroacoustic applications. Indeed, they enable designers to relate precisely emitted aerodynamic noise[9] to the dynamics of coherent vortices and to define appropriate noise-control strategies in terms of vortex manipulation. We recall that coherent vortices are low-pressure regions and hence potential sources of acoustic waves (which are pressure waves). Let us mention finally the LES of a flow within a cavity carried out by Zang et al. [297] using the dynamic mixed model. In this simulation, the flow is forced above by a moving lid, and good experimental agreement is obtained.

Spanwise groove

We briefly review now a LES of a flow in a plane channel at Mach 0.3 carried out by Dubief [77] (see also Dubief and Delcayre [78]). One side of the channel is flat, and the other is equipped with two small, spanwise, square grooves of size 40 flat-wall units. Periodicity is assumed in the streamwise

[9] The frequency spectrum of the acoustic power may easily be determined from the pressure spectrum (see, e.g., Mankbadi [195]).

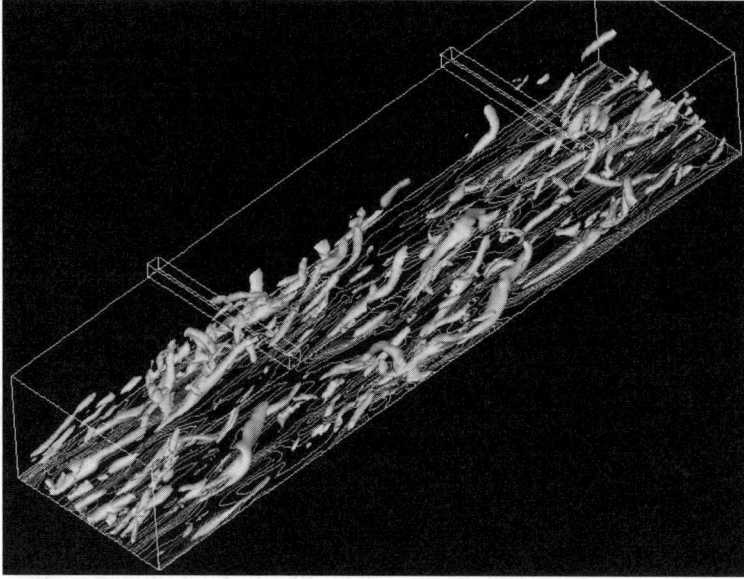

Figure 7.26. Positive Q isosurfaces on the flat plate of a Mach 0.3 periodic channel equipped with small spanwise grooves on the other side. (From Dubief and Delcayre [78]; courtesy Y. Dubief and *Journal of Turbulence*.)

and spanwise directions. The Reynolds number h^+ on the flat plate is 160, and the orthogonal grid is stretched in the streamwise and spanwise directions with a minimum grid spacing (in flat-wall units) of, respectively, 2 and 0.4 in these two directions. Hence, this is a very well resolved simulation close to the wall. The number of grid points is $200 \times 128 \times 64$. At this low Mach number, compressibility effects are negligible, and it was checked by Dubief [77] that the flat-wall velocity statistics were in very good agreement with Lamballais's [151] DNS results in an incompressible channel between two flat walls at the same Reynolds number $h^+ = 160$. Lamballais's DNS studies were carried out with a very precise code, combining pseudo-spectral methods in the directions of periodicity and compact finite differences of sixth order in the direction perpendicular to the wall (see Chapter 5). The accuracy of his code has been checked to be close to spectral in such a case by extremely good comparisons with purely spectral DNSs (see Chapter 5). This constitutes an excellent validation for the compressible LES code, which involves about 1.6 million grid points, and is not very expensive. In fact, the LESs of a grooved channel show that the boundary layer on a grooved wall is slightly affected (see Dubief [77] for more details). Animations of Q on the flat and grooved walls presented on the CD-ROM (Animation 7-3) as well as in Dubief and Delcayre [78] permit numerous "semi hairpins" traveling downstream to follow. Their legs form quasi-longitudinal vortices close to the wall of approximate length 300 wall units, and many of them have a self-induced, raised-arch-form tip. Figure 7.26 shows such a field on the flat side of the channel. Animation 7-3 gives more detailed information on Q isosurfaces and ω_z isolines at the walls

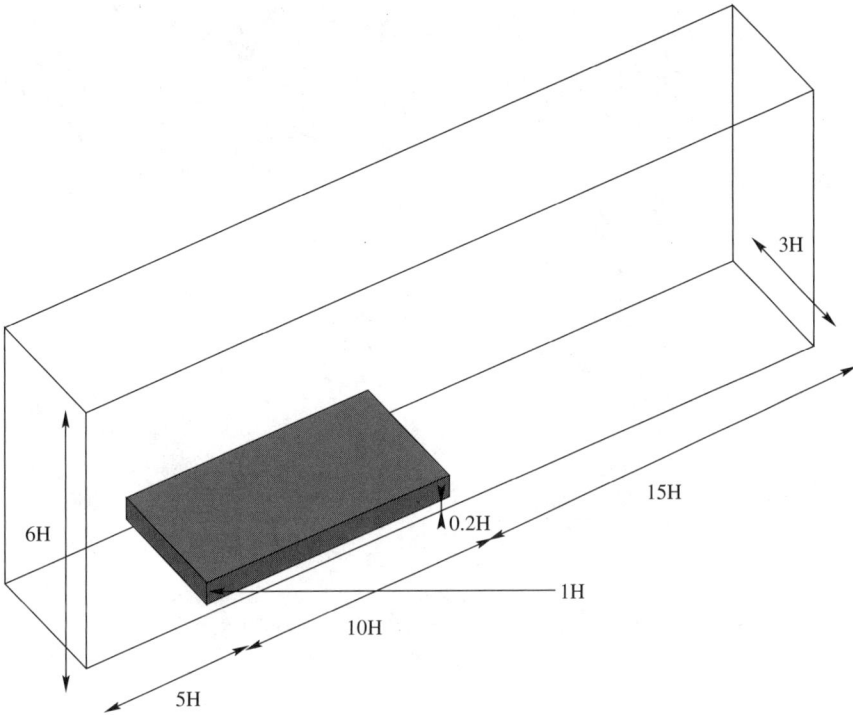

Figure 7.27. Rectangular obstacle with a wall effect.

with perspective and top views of the flow close to the flat and ribbed walls. For the spatial boundary layers previously presented, high values of ω_z at the wall (in red) correspond to regions of high friction and should be correlated with the high-speed streaks.

Briand [31] has identified the same type of vortices in the LES of a spatially developing turbulent boundary layer at Mach 0.3 already presented and has checked that they travel at the mean-flow velocity corresponding to the location of their tip, which is in agreement with the experimental observations of Stanislas [275] and Adrian and collaborators [45]. Let us mention finally the numerous experimental and numerical results on boundary layers and channel-flow structure provided in the very well documented book of Bernard and Wallace [22].

7.3.3 Obstacle with a wall effect

We present now LES of a weakly compressible flow at Mach 0.2 around a two-dimensional (infinite in the spanwise direction) rectangular obstacle of thickness H and length $10H$ that lies parallel to a wall and is located at a height of $0.2H$ above it. This work is reported in Lesieur et al. [176]. Periodicity is assumed in the spanwise direction. The Reynolds number based on the velocity at infinity U_0 and H is 165,000. The geometric configuration is presented in Figure 7.27 with flow going from left to right. The spanwise

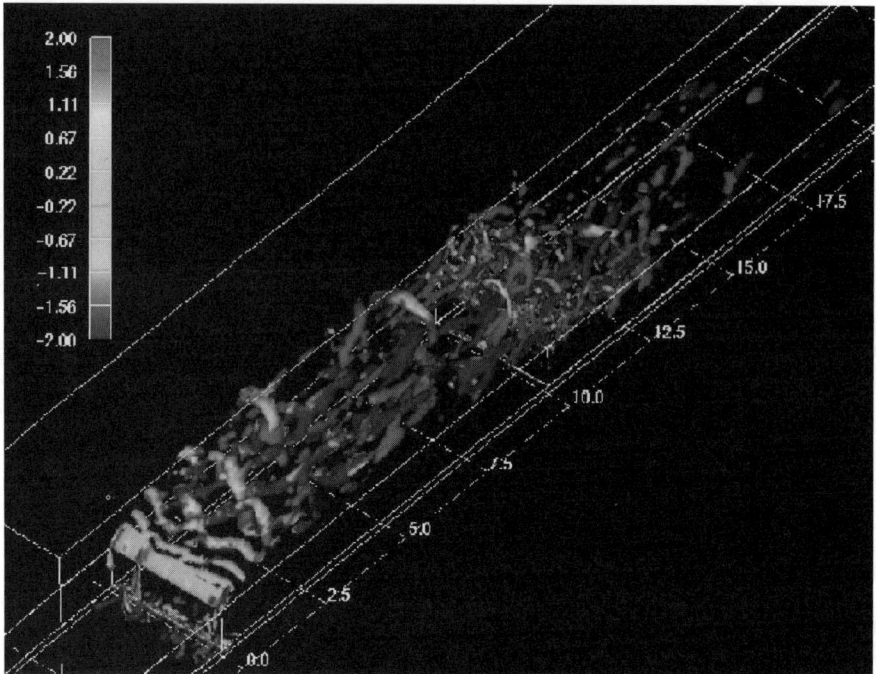

Figure 7.28. Positive Q isosurfaces around a rectangular flat plate with a wall effect. Positive and negative longitudinal vorticity is also shown. (Courtesy P. Begou.)

width of the domain is $3H$. The upstream velocity is identical to the case of the large, two-dimensional square cavity already studied with $\delta = 0.1H$. A grid of 1,542,000 points split into four subdomains is used with the first point above the upper wall being located at a distance of 15 wall units. Figure 7.28 and Animation 7-4 show the main features of this complex flow very well. Upstream, most of the fluid rises against the step, except for a small fraction that passes in the channel between the wall and the obstacle. Then the flow detaches above the forward ridge, forming an unstable vortex sheet that rolls up into two-dimensional KH vortices, which are then shed and seem to be destabilized under the action of pairs of hairpin vortices impinging on the rising step. The flow reattaches on the upper wall, whereas KH vortices transform into arch vortices similar to those already obtained in various situations in which detachment and reattachment are involved. These arch vortices detach behind the rear backward-facing step. The flow in this region is interesting because a rising current coming from the local recirculating bubble hits the current passing under the obstacle. Further downstream, the arch vortices weaken substantially while exiting the domain.

Animation 7-4 presents isosurfaces of positive Q in green, positive longitudinal vorticity in red, and negative longitudinal vorticity in blue as the viewer rotates with respect to the flow. Initially, the flow goes from left to right, then one sees it in perspective from upstream, and finally it goes from right to left.

Figure 7.29. Dennis Conner's *Stars and Stripes* training in San Diego. (Courtesy C. Agnus and J. Lesieur (L'Express) and Grenoble-Sciences [167].)

7.3.4 Pipe flow

There is currently much less LES work done on pipes than on plane channels because pipes are more complex geometrically. We briefly discuss here the work of Nicoud and Ducros [219] using unstructured grids. The Reynolds number based on the diameter $2R$ and the bulk velocity (see exact definition to follow) is 10,000 (with $R^+ = 320$), and the Mach number is 0.25. Nicoud and Ducros use the so-called wall-adapting local eddy viscosity (WALE) subgrid model (see their paper for more details), where the eddy viscosity has the correct scaling $O(y^3)$ close to the wall. In a former work, Ducros et al. [82] applied the filtered Smagorinsky model[10] to the pipe. Both studies show good results in terms of statistics. They also display the same low-and high-speed streaks and quasi-longitudinal vortices as found in boundary layers and channels on a flat plate.

7.4 Drag reduction by riblets

Let us briefly recall the numerous studies associated with passive turbulence control by longitudinal riblets put on some parts of planes, boats, and more recently on competition swimsuits made of so-called sharkskin. Just before Dennis Conner won the 1986 final of the America's Cup against the Australians (see Figure 7.29), he spoke of a secret weapon he had, and his boat

---

[10] This is the equivalent of the FSF model [81] for the Smagorinsky model.

Stars and Stripes was hidden from the public. After the victory, he revealed that riblets had been installed on the hull of the boat. This made headlines in newspapers worldwide. Meanwhile, defense agencies in several countries were sponsoring classified research on riblets and were testing riblet-equipped planes.

The optimal spanwise wavelength of triangular riblets was empirically found to be $\lambda_z^+ = 10\text{--}20$, which is approximately 10^{-5} m in air and 10^{-4} m in water. This determination requires knowledge of the ratio v_*/U_∞ in terms of the Reynolds number, for which empirical laws exist. In fact, the DNS studies of Choi et al. [40] using equilateral triangles have shed new light on the role of quasi-longitudinal vortices in drag reduction by riblets. Indeed, the longitudinal vortex diameter is $d^+ \approx 25$. Choi et al.'s simulations show that for λ_z^+ larger than 25 (they took 40), the quasi-longitudinal vortices are trapped in the valleys of the riblets, which increases the drag. However, in the simulation with $\lambda_z^+ = 20$, the longitudinal vortices sit above the riblets' peak, and the drag is decreased. This may be interpreted in the first case ($\lambda_z^+ = 40$) by an increased effective (in terms of drag reduction) contact area of the fluid with the wall. In the second case ($\lambda_z^+ = 20$), the effective contact area is reduced.

A very important question for aeronautic applications concerns the influence of compressibility in a perfect gas for riblet efficiency. Because riblet size scales in the incompressible case with wall units, it is interesting first to ask how the spanwise wavelength of streaks and the vortex diameter (respectively, 100 and 25 viscous units) are modified by compressibility. In fact two DNSs of a compressible flat-wall channel at Mach 1.5 and a Reynolds number of 3,000 have been carried out by Coleman et al. [48] (using spectral methods) and Lechner et al. [160] (using high-order finite differences). In these compressible channel calculations, the bulk density ρ_b and velocity U_b are defined by

$$2h\rho_b = \int_{-h}^{+h} \langle \rho \rangle dy, \quad 2h\rho_b U_b = \int_{-h}^{+h} \langle \rho u \rangle dy. \tag{7.51}$$

These calculations are carried out at fixed bulk density and wall temperature T_w whatever the Mach number. The latter is defined as U_b/c_w, where $c_w = \sqrt{\gamma R T_w}$ is the sound speed at the wall. The Reynolds number is $\rho_b U_b h / \mu_w$, where μ_w is the dynamic viscosity at the wall. For each U_b, the simulation is done at constant mass flux to generate a turbulent state rapidly. The velocity gradients within the channel produce a heating by molecular diffusion, and the channel interior becomes warmer than the walls. Coleman et al. [48] and Lechner et al. [160] show that, when turbulence has developed, the average temperature (resp. density) remains approximately uniform in the major part of the channel but decreases (resp. rising) close to the walls. The inner plateau part of the density is very slightly less than ρ_b. The low- and high-speed streaks are more elongated, but their spanwise size in physical units (not wall units) remains unchanged. Vortices are also very similar to the plots of Figure 7.26.

Let us consider now how some statistics of the flow are modified by compressibility. Let ρ_w be the average density on the wall. We recall that the wall unit is given by

$$l_v = \frac{\nu_w}{v_*} = \frac{\mu_w}{\rho_w v_*} = \frac{\mu_w v_*}{\tau_w},$$ (7.52)

where

$$\tau_w = \rho_w v_*^2 = \mu_w \frac{\partial u}{dy}\big|_w$$ (7.53)

is the stress at the wall. This leads to

$$l_v = \frac{\mu_w}{\sqrt{\rho_w \tau_w}}.$$ (7.54)

Because μ depends only on temperature,[11] μ_w will be time invariant. However, recent LESs of the same problem using the SSF model and well-validated immersed-boundary methods have been carried out by Hauët [118], who found that the stress at the wall τ_w is approximately unchanged up to Mach 1.5. Assuming this constant stress gives $l_v \propto \rho_w^{-1/2}$. Because ρ_w increases with Mach number, l_v will decrease. This fact has been checked in the aforementioned simulations in which h^+ does increase with the Mach number. Looking at the spanwise autocorrelation of the longitudinal velocity, Hauët [118] finds that spanwise wavelength (in physical units) of the streaks is increased by approximately 40% with respect to a Mach 0.33 simulation, which is at variance with the conclusions of Coleman et al. [48] and Lechner et al. [160]. More research is needed to clarify this point, for immersed-boundary techniques are quite questionable owing to uncertainties associated with the velocity of fictitious flows contained within obstacles. Further uncertainties exist in the compressible case. However, Hauët's code has been validated by simulating properly a laminar flow around an infinitely thin heated plate located at the center of a channel.

Hauët [118] has also developed a LES of a compressible channel, one side of which is equipped with longitudinal, triangular riblets. Two riblets were studied: the "high" one, with height and width (at Mach 0.33), respectively, of 11 and 22 wall units, and the "great" one, with height and width (at Mach 0.33), respectively, of 22 and 44 wall units. Hauët has first validated satisfactorily at low Mach the numerical code used against Choi et al.'s

---

[11] The dependence is through either the Sutherland law or the $\mu \propto T^{0.7}$ law taken by Coleman et al. [48] and Lechner et al. [160].

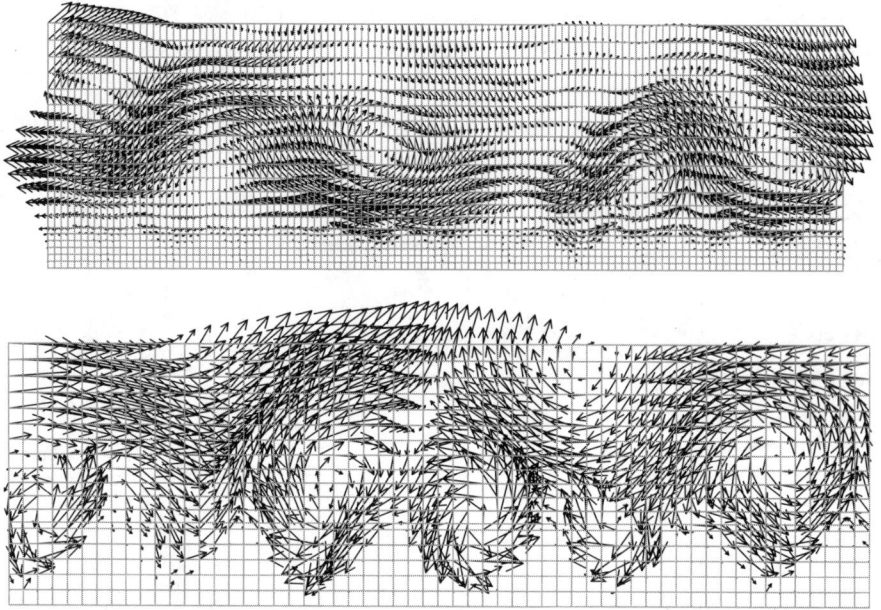

Figure 7.30. Cross section of the velocity in a channel above riblets at Mach 0.33. (Top) "High" riblet; (bottom) "great" riblet. (Courtesy G. Hauët.)

DNS. The physical size of each system of riblets was unchanged when going from Mach 0.33 to Mach 1.5. In these simulations, the high riblet turns out to reduce the drag ($\approx$5% for the mean friction coefficient at Mach 1.5, versus $\approx$3% at Mach 0.33). The "great" one increases it. Hauët [118] recovers the same vortex phenomenology as Choi et al. [40], with longitudinal vortices above the riblet tips in the high-riblet case and inside the valleys in the great-riblet case. This is confirmed by Animations 7-5 and 7-6, which present longitudinal vorticity in Hauët's high and great riblets at Mach 0.33. Figure 7.30 shows an instantaneous projection of the velocity vector in a cross section for the two riblets. It is clear from these plots that alternate vortices lie within the valleys for the great riblet, whereas thinner longitudinal vortices stay above the peaks for the high riblet. Corresponding animations of cross sections of low- and high-speed streaks by a plane located 10 wall units away from the riblet peaks are also presented on the CD-ROM (Animations 7-7 and 7-8). The straight lines indicate positions of riblet valleys. The streaks are much more coherent longitudinally for the high riblet than for the great one.

If, in a free, compressible boundary layer, the optimal physical size of riblets does not vary from subsonic to supersonic regimes, then an airplane with the same riblet system will be able to reduce drag at all speeds. Similar conclusions have been drawn from experiments carried out at ONERA-Toulouse by Coustols and Cousteix [57]. This is quite satisfactory from the point of view of aircraft designers.

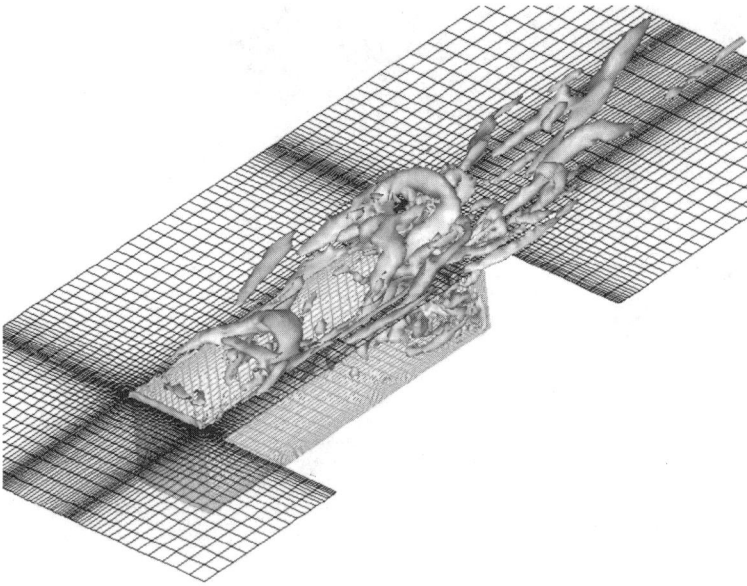

Figure 7.31. Perspective side view of positive Q isosurfaces in a transonic flow on a rectangular cavity. (From Dubief and Delcayre [78]; courtesy Y. Dubief.)

7.5 Transonic flow past a rectangular cavity

We show now the LES results of a flow at Mach 0.91 past a rectangular-parallelepiped cavity, also presented in Dubief and Delcayre [78]. The flow corresponds to an experiment of Tracy and Plentovich [281]. Let H be the cavity depth. The Reynolds number based on the free-stream velocity and H is equal to 1.25×10^6, and the 99% boundary layer thickness upstream of the cavity is $0.3\,H$. The inflow is a mean profile perturbed by three-dimensional white noise. The resolution is coarse because the first point away from the wall is at a distance of 70 upstream wall units. More specifically, there are $50 \times 30 \times 30$ points inside the cavity, and $100 \times 40 \times 70$ above. No wall law is used. The Q-criterion has also been applied in this case: Animation 7-9 (presented on the CD-ROM and in Dubief and Delcayre [78]) shows how large Λ vortices shed behind the upstream backstep travel with the mean flow, impinge on the ridge of the upstep, and are carried away downstream. This is clear from Figure 7.31 (taken from Dubief [77]). There is also an important recirculation in the rear part of the cavity. Although the upstream turbulent boundary layer is not properly described, pressure spectra accurately predict the fundamental frequency of the vortex shedding. Two other low frequencies have also been identified and are in good agreement with the experiment of Tracy and Plentovich [281] within the convergence error of the spectra for the lowest frequency. The level of acoustic noise emitted above the cavity in the computation of Dubief is overestimated by 10–20 dB because of the confinement of

Figure 7.32. View of a preliminary model of *Hermès*. (Courtesy Dassault and Grenoble-Sciences [167].)

the domain and open boundary conditions. A last question that arises here concerns the validity of the Q-criterion for strongly compressible flows. Although theoretical considerations justifying this criterion (in terms of local pressure minima at least) assume incompressibility, it seems here, from the good agreement with experiments as far as Strouhal numbers of the shed Q vortices are concerned, that the criterion still works. The Q-criterion has also been applied with success by Lechner et al. [160] for the previously mentioned Mach 1.5 channel DNS.

7.6 European space shuttle *Hermès*

Let us present a LES done by David [62] of the detached boundary layer over a curved compression ramp at Mach 2.5 modeling the wind-side region of the body flap of the European space shuttle *Hermès* (see Figure 7.32) during its projected reentry.[12] The external Mach number relevant to the shuttle is about 10 (at altitude 50 km, incidence angle 30°, and flap extension angle $\alpha_0 = 20°$). The whole computational domain is contained within the bow shock. The grid used is shown, upside down, in Figure 7.33. The resolution is

---

[12] Since then, the *Hermès* project has unfortunately been canceled because of budget cuts to the program.

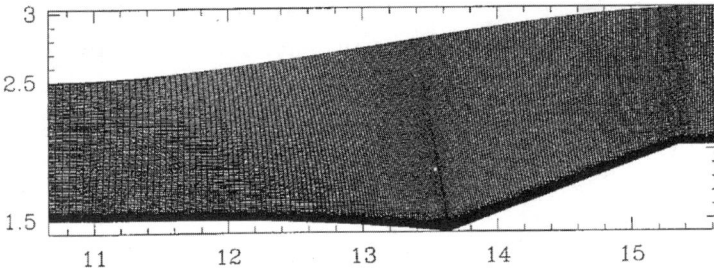

Figure 7.33. Transverse section of the 220- × 140- × 25-point grid used for the simulation of the transition on the curved ramp (angle 20°). The axes are graded in meters counted from the nose of the full-size shuttle. The spanwise size of the domain is 4.5 times the displacement thickness δ_i of the upstream boundary.

$220 \times 140 \times 25 = 770,000$ grid points. The first part of the boundary (up to 13.6 m away from the nose) is curved. It corresponds to the wind side of the body. The ramp is the body flap, which is assumed to be flat. For computational reasons, it is extended by a fictitious horizontal surface introducing a cutoff with the lee side of the flap and the afterbody. This enables the prescription of well-posed boundary conditions at the exit of the domain. The simulation requires knowledge of the density, temperature, and velocity profiles at the upstream boundary of the domain. Because these are not available for in-flight conditions, we simulate a well-documented 1/90 experiment performed at ONERA in the wind tunnel R3CH. Our upstream condition results from the experimental profiles plotted in Figure 7.34 with white noise of amplitude $(2 \times 10^{-3}) U_\infty$ superimposed on the three components of the velocity at each time step. On the model, the wall temperature is $T_w = 290$ K, and the "external" (outside of the boundary layer but inside the bow shock) temperature is $T_\infty = 460$ K. Let us recall the adiabatic temperature T_a given in Chapter 1 by Eq. (1.17) defined as the temperature reached at the wall (where the velocity is zero) by a fluid parcel traveling adiabatically from the exterior of the boundary layer (for a time-independent perfect fluid). Equation (1.17) can be written as

$$T_a = T_\infty \left(1 + \frac{\gamma - 1}{2} M_\infty^2\right). \tag{7.55}$$

We have $T_a = 1,035$ K for $M_\infty = 2.5$ and $\gamma = 1.4$, yielding $T_w / T_a = 0.28$. The ramp is therefore very cool with respect to the ambient fluid, which models the radiative balance of the true shuttle during its reentry.

The measured upstream displacement thickness of the boundary layer is $\delta_i = 0.21 \times 10^{-3}$ m, yielding a Reynolds number $Re_{\delta_i} = 727$. This is too high for the code described here. The simulation is therefore performed at the maximal Reynolds number permitted by our resolution, which is $Re_{\delta_i} = 280$. For this reason, the results presented in the following have to be considered as qualitative only. One should also bear in mind that the similitude between

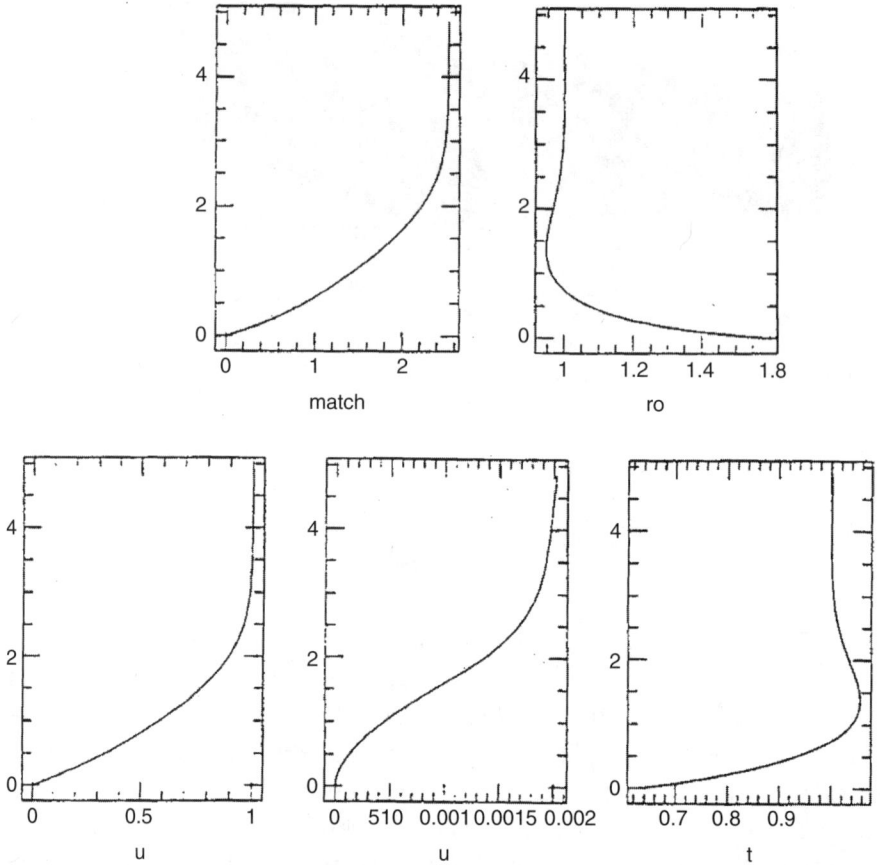

Figure 7.34. Profiles prescribed at the upstream boundary (Mach number and normalized density ρ/ρ_∞ on top; streamwise and transverse normalized velocity components u/U_∞ and v/U_∞ and normalized temperature T/T_∞ at the bottom). For all plots, the vertical co-ordinate is ξ_2/δ_i. For the 1/90 experiment, $\rho_\infty = 7.685 \times 10^{-2}$ kg/m³, $U_\infty = 1,089$ m/s, and $T_\infty = 460.3$ K.

the experiment and the in-flight conditions cannot be exact. If the Mach and Stanton numbers are in similitude, it is extremely unlikely that the Reynolds numbers also are. Figure 7.35 shows the detachment of the boundary layer and its reattachment to the flap obtained from a preliminary two-dimensional simulation. One sees clearly the multiple-legged Λ shock focalizing outside of the domain. Its position fluctuates in time, owing to the large vortices in the recirculation zone around the hinge. However, the most interesting feature of the flow is not reproduced in this two-dimensional simulation: Between its detachment and reattachment, the boundary layer undergoes a certain curvature, whose radius R can be roughly estimated from Figure 7.35. This yields a Görtler number

$$\mathcal{G} = Re_{\delta_i}\sqrt{\frac{\delta_i}{R}} \approx 2\text{--}3,\tag{7.56}$$

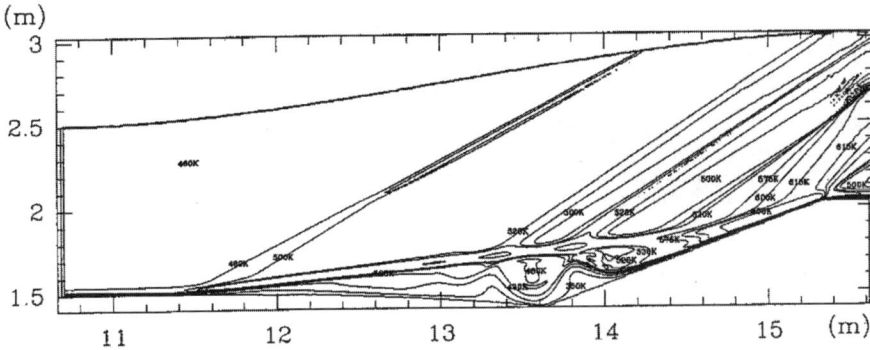

Figure 7.35. Instantaneous temperature map obtained from a preliminary two-dimensional simulation of the flow over a curved ramp. Here again, the axes correspond to the full-size shuttle, whereas it is the 1/90 experiment that is actually simulated.

which is high enough to give rise to centrifugal instability according to linear stability theory. Experimental evidence of streamwise counterrotating Görtler vortices in a similar case was produced by Settles et al. [261], but the consequence of these vortices on the wall heat flux has remained an open question. Figure 7.36 shows such Görtler vortices obtained from the three-dimensional LES performed by David [61] using the SSF model in a domain of spanwise extension equal to $4.5\delta_i$. One clearly sees two large structures, crosscuts of which show that each of them corresponds to a pair of counterrotating Görtler vortices. These crosscuts in a plane perpendicular to the flap and located

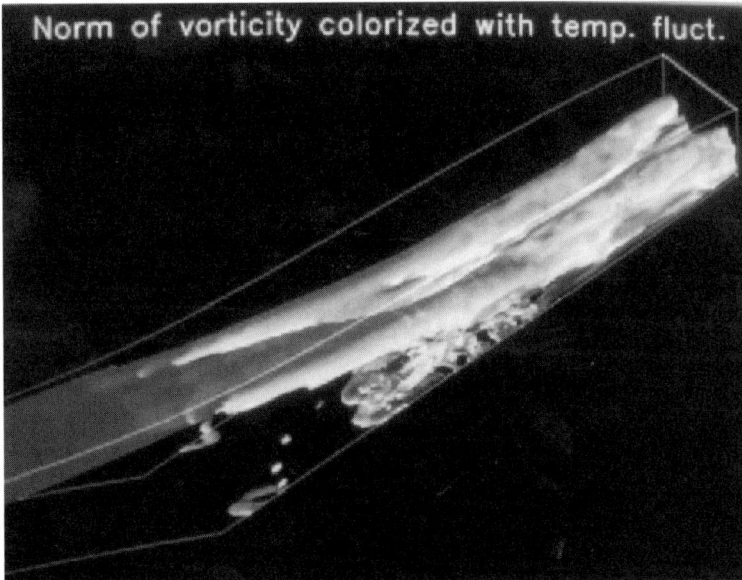

Figure 7.36. Ramp flow. Close-up of the hinge and body-flap region showing an isosurface of the vorticity magnitude. This surface is shadowed by temperature. (Courtesy E. David [61] and Grenoble-Sciences [167].)

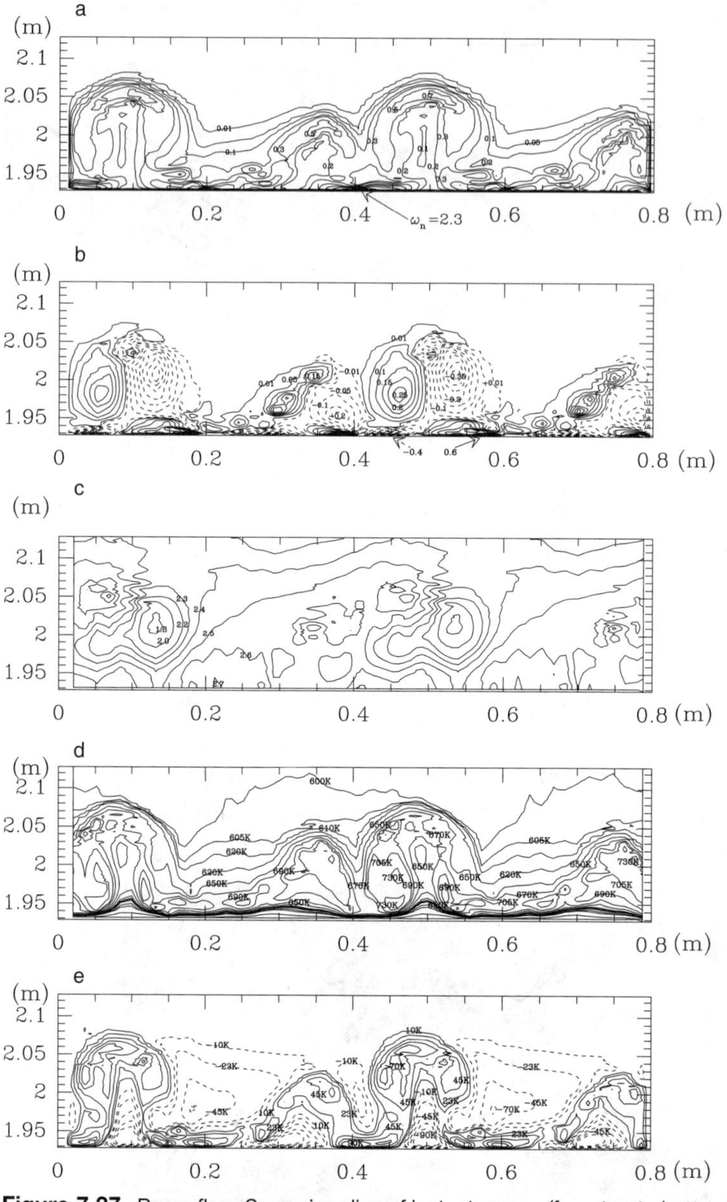

Figure 7.37. Ramp flow. Spanwise slice of instantaneous (from top to bottom) vorticity modulus, longitudinal vorticity, pressure, temperature, and temperature fluctuations. Dashed lines correspond to negative values. Graduations are again relevant to the full-size shuttle.

approximately 15 m downstream of the nose are presented in Figure 7.37, which displays the instantaneous vorticity modulus, longitudinal vorticity, pressure, temperature, and temperature fluctuations with respect to a spanwise average. In this figure, the slice is repeated twice in the spanwise directions. This is permitted by the periodic boundary conditions and makes the vortex structure easier to understand. The longitudinal vorticity plot shows

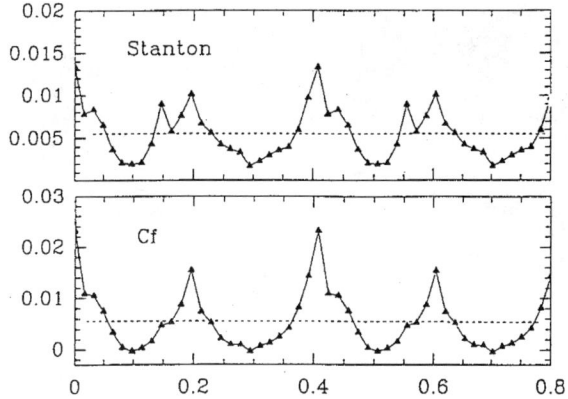

Figure 7.38. Instantaneous profiles of Stanton number St (top) and skin-friction coefficient C_f (bottom).

that the big vorticity dipole located between the spanwise locations $\xi_3 = 0$ and $\xi_3 = 0.2$ m (coordinates relevant to the full-size shuttle; they have to be divided by 90 to correspond to what is actually simulated, i.e., the model) is made of one cyclonic (anticlockwise) vortex, one longitudinal vortex between 0 and 0.1, and one anticyclonic (clockwise) vortex between 0.1 and slightly less than 0.2. Under the latter, a flat region of antivorticity forms at the wall. The second dipole is smaller. Interestingly, its cyclonic branch consists of two small vortices that seem to pair. Under the anticyclonic branch, antivorticity exists also. There is a very good correlation between the vorticity and temperature structures. A first remark is that the flow close to the wall is very hot with temperatures that may reach 730 K (i.e., 70% of the adiabatic temperature) at a spanwise distance ξ_3 slightly larger than 0.4 m (i.e., zero due to spanwise periodicity). At this point, it is clear that the longitudinal cyclonic vortex produces a downwashing to the wall of higher external fluid, which was originally at a temperature of 600 K. During the process, the temperature significantly increases as one approaches the wall. In fact, the presence of these pairs of counterrotating Görtler vortices allows us to understand Figures 7.37 and 7.38. The latter represents the corresponding instantaneous profiles of the Stanton number St and skin-friction coefficient C_f. Uplift of slow and cold fluid from the boundary, which occurs in between each pair of counterrotating vortices, implies negative temperature fluctuations and minima of C_f and St. The latter are located at $\xi_3 = 0.1$ m $\equiv 0.5$ m with a secondary minimum at 0.3 m. Conversely, maximal values of C_f and St are found at $\xi_3 = 0.2$ m and 0.4 m, which is half way between the two pairs, where the downwash of hot (and fast) fluid from the outer part of the layer is maximal. The extreme values of the temperature fluctuations (with respect to the time average) plotted in Figure 7.37(e) are ± 90 K. They are found close to the wall (which is at $T_w = 290$ K). These 30% of the temperature fluctuations induce huge fluctuations of the Stanton number (Figure 7.38), between 2×10^{-3} and 14×10^{-3} with an average of about 6×10^{-3}. The rms Stanton number is thus 133%.

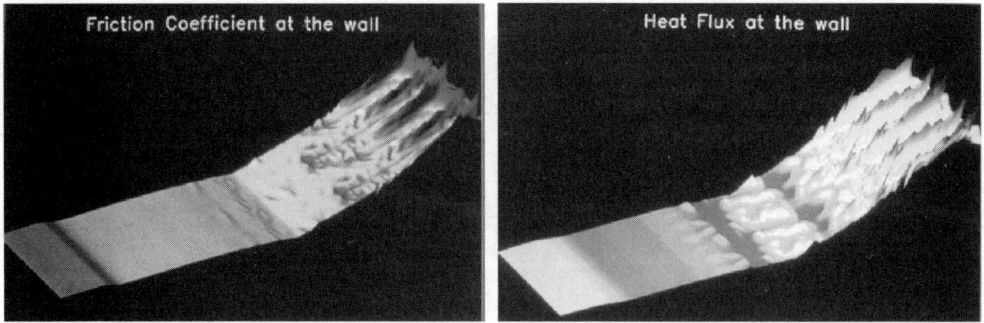

Figure 7.39. Instantaneous contours (with elevation) of the skin-friction coefficient C_f (left) and the Stanton number St (right).

The same trend is observed for the skin-friction coefficient C_f displayed in Figure 7.38, which remains approximately proportional to St as predicted by the strong Reynolds analogy. An analogy factor $\bar{s} = \overline{St}/2\overline{C_f} \approx 2.9$ can be (quite roughly) estimated from the mean values of St and C_f. In trying to work out analogy factors associated with the peak and valleys of C_f and St, one finds $s_{max} = 1.1$ and $s_{min} \longrightarrow \infty$, respectively (because C_f goes to zero). This clearly shows that the strong Reynolds analogy, although globally satisfied, cannot be relied upon to deduce local Stanton numbers out of local values of the skin-friction coefficient. Finally, the elevated contour maps of C_f and St shown in Figure 7.39 prove that these values – recorded from an instantaneous cross section of the flow – are almost independent of the streamwise coordinate ξ_1. Time-averaged plots (not shown here) also prove that the Görtler vortices are, in this simulation, fairly stable in time. This is likely to enhance their destructive effects considerably on the material of the body flap. A question that will remain a mystery is whether the presence of these vortices on the real *Hermès* would really have induced temperature fluctuations capable of destroying the rear flap. These results are quite pessimistic in this context. Indeed, Eq. (7.55) shows that $T_a/T_\infty = 2.25$ for $M_\infty = 2.5$. If one admits that in reality the maximum temperature at the wall will be $\approx 0.7T_a = 1.57T_\infty$, and if one takes $T_\infty = 3,000$ K, the flap would be in contact with fluid at temperatures of the order of $4,700$ K.

A last important remark for modeling strategies is that such vortices are not predicted by industrial numerical models used for the design.

7.7 Heat exchanges in ducts

For numerous applications, particularly those of engineering interest, it is important to reach a deeper understanding, and to be better able to predict, the heat exchanges between a heated wall and the surrounding turbulent flow. This is relevant if, for example, one wants to improve the performances of a heat exchanger or the cooling of a rocket engine.[13] For cooling purposes, the cold fluid often flows in ducts of square or rectangular cross section and exchanges heat with the hot fluid through one of the walls. Flows in square and rectangular ducts are characterized by the existence of secondary flows, called Prandtl's flow of the second kind, driven by the turbulent motion and consisting of a mean flow perpendicular to the main flow direction. Their intensity is 1–3% of the mean streamwise velocity, but their effect on heat and momentum mixing is quite significant. To reproduce this weak secondary flow properly with a RANS approach, elaborate second-order models have to be employed. Furthermore, RANS heavily relies on empirical models to represent the near-wall dynamic and thermal behavior. We will show here that LES provides an excellent tool for correctly reproducing the heat exchanges in closed ducts. The reader will find more details in Métais [207]. The LES studies presented here use the SSF model.

7.7.1 Straight ducts of square section

We are going to present LES of turbulent flow within a square duct on the basis of the work of Salinas and co-workers [249–251]. We again use the Grenoble COMPRESS code. Let $Re_b = 6,000$ be the Reynolds number based on the bulk velocity. We consider successively the isothermal case (with the four walls at the same temperature) and the heated duct (for which the temperature of one of the walls is set higher than the temperature of the three other walls). Moderate resolutions are used because the grid has $32 \times 50 \times 50$ nodes in the isothermal case and $64 \times 50 \times 50$ nodes in the heated case along x (streamwise), y, and z (transverse) directions. This renders the computation very economical with respect to a DNS. To correctly simulate the near-wall regions, a nonuniform (orthogonal) grid with a hyperbolic-tangent stretching is used in the y and z directions. The minimal spacing near the walls is 1.8 wall units. The Mach number based on the bulk velocity and the wall temperature is $M = 0.5$. Imposing a uniform temperature at the walls is compatible with the use of periodic boundary conditions in the streamwise direction, which simplifies the computation.

Isothermal case

The LES has been validated against the incompressible DNS of Gavrilakis [107], and very good agreement has been obtained. The secondary flow reveals the existence of two streamwise, counterrotating vortices in each corner of the duct. The maximum velocity associated with this flow is 1.169% of the bulk velocity. This agrees very well with experimental measurements. It shows the

---

[13] The *Ariane V* Vulcain engine is an example.

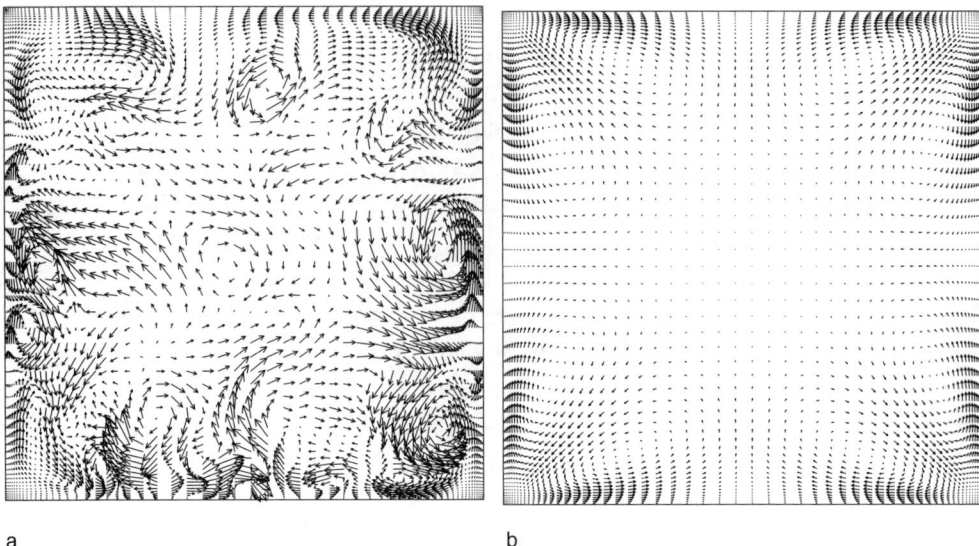

a   b

Figure 7.40. Velocity vector in a cross section of an isothermal duct. (a) Instantaneous field; (b) mean field. (Courtesy C. Salinas-Vazquez)

ability of LES to accurately reproduce statistical quantities. Figures 7.40(a) and 7.40(b) show a projection of, respectively, the instantaneous and average flow field in the duct cross section. It clearly indicates a very pronounced flow variability with an instantaneous field very distinct from the mean field. The maximum for the transverse fluctuating velocity field is of the order of ten times the maximum for the corresponding mean velocity field.

Heated wall case

We consider now a square duct subjected to an asymmetric heat flux. Let T_h be the temperature of the hot wall. Salinas-Vazquez and Métais ([250, 251]) have studied the effect of varying the temperature ratio between the hot wall and the other walls. When the heating was increased, they observed an amplification of the mechanism of ejection of hot fluid from the heated wall. Figure 7.41 shows velocity and temperature maps near the heated wall (only one portion of the duct is represented). We can see that these ejections are concentrated near the middle plane of the heated wall. This yields a strong intensification of the secondary flow. We can check also that the turbulent intensity is reduced near the heated wall because of an increase of molecular-viscous dissipation in that region caused by the strong heating.[14]

Let us consider the mean heat flux

$$q_w = [\langle \kappa \rangle \partial \langle T \rangle / \partial n]_w, \tag{7.57}$$

---

[14] Here, dynamic molecular viscosity μ increases as a result of the heating, and density decreases; thus, kinematic viscosity ν will increase also.

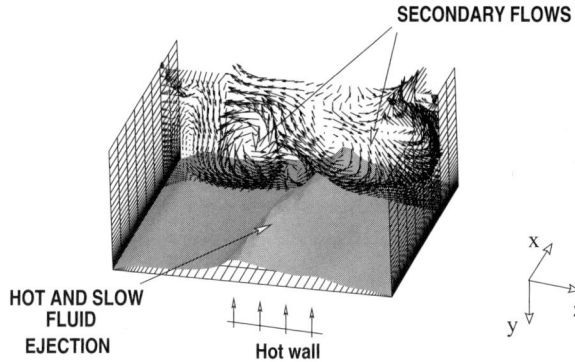

Figure 7.41. Large-scale motion over the hot wall in a heated duct ($T_h/T_w = 2.5$) showing instantaneous velocity vector in a cross section and isosurface of temperature ($T/T_w = 2.1$). (Courtesy C. Salinas-Vazquez.)

where n is the direction normal to the wall. Brackets stand for averaging in the flow direction x and in time. We have seen that, near the duct middle plane, slow and hot fluid is ejected from the wall toward the duct interior. This induces a strong reduction of the longitudinal velocity gradient $\partial U/\partial n$ in that region in the strongly heated cases and a reduction of the wall shear stress

$$\tau_w = [\langle\mu\rangle\partial\langle U\rangle/\partial n]_w \qquad (7.58)$$

because the velocity-gradient decrease overwhelms the viscosity increase. Similarly, the temperature gradient normal to the heated wall is significantly reduced when the heating is strong enough to yield a unique violent ejection concentrated in the middle of the heated wall (see [251]). Recalling that the molecular Prandtl number is a constant, we see that the mean heat flux is also reduced here. Outside the middle plane, the reinforcement of the secondary flows with heating is accompanied by a stronger impingement of the heated wall by the fluid coming from the duct core. It generates more significant velocity gradients and greater wall shear stress at a distance of about 0.2 hydraulic diameters from the lateral wall. This effect is not so clearly marked for the heat flux, and the heat flux from the hot wall is globally reduced in the strongly heated case mainly because of the local flux decrease associated with the central plane ejection.

7.7.2 Ducts with riblets

In a recent study performed by Issa [132], longitudinal ribs were put on the heated wall to prevent the formation of this ejection and therefore to increase the heat flux given by Eq. (7.57). Two ribs were placed symmetrically with respect to the heated wall middle plane, and various shapes of ribs were tested. Figure 7.42 displays a velocity cross section for the duct both without and with ribs. Three shapes (called circular, triangular, and square) of ribs are studied. Recall that heating produces an intensification of the secondary mean flow.

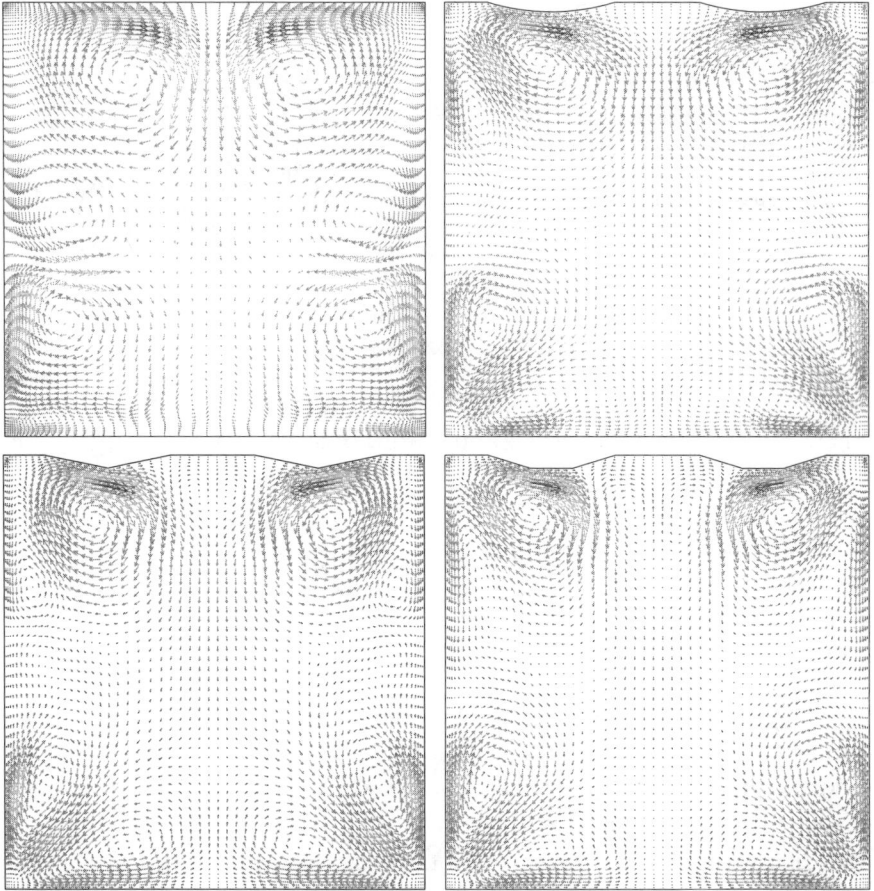

Figure 7.42. Mean velocity vector projected on a cross section of the duct. (Top left) square duct without ribs; (top right) square duct with circular ribs; (bottom left) square duct with triangular ribs; (bottom right) square duct with square ribs. The heated wall is located at the top of each section. (Courtesy R. Issa.)

When ribs are present, these vortices sit close to the top of the rib but on the side close to the lateral wall. (They are rounder with the triangular and square ribs.) Hence they are more distant than in the heated flat-wall case, reducing their mutual interaction, which results in less hot fluid being pumped away from the wall. Thus, the temperature gradient at the wall and the mean heat flux are intensified by as much as 15% for some shapes of ribs.

7.7.3 Spatially growing turbulence through a straight duct

In Salinas-Vazquez and Métais's [251] work, square ducts with a prescribed temperature at each wall were considered. These boundary conditions for the temperature are compatible with periodic boundary conditions in the

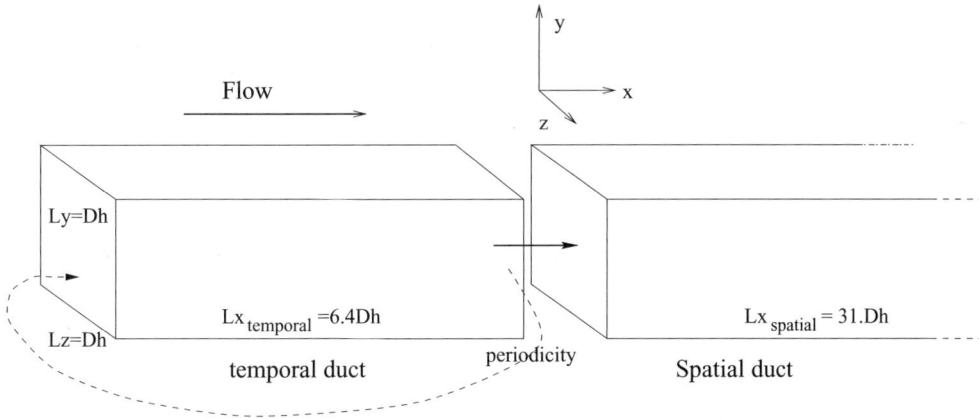

Figure 7.43. Numerical upstream configuration for the spatially developing duct.

streamwise direction. However, in many industrial applications, a given heat flux is actually imposed at the heated wall and the turbulent flow is no longer periodic in the x direction, owing to the continuous energy increase along this direction. One then has to deal with spatially developing turbulent flows, which require a more complex prescription of inflow and outflow boundary conditions (see [240]). The Mach number is still 0.5 and the bulk-velocity-based Reynolds number is 6,000. Fully turbulent inlet boundary conditions are provided at each time step by a LES of a duct with all walls at temperature T_w. This longitudinally periodic duct (called a temporal duct) is linked to the spatially growing duct through the characteristic boundary conditions proposed in [240] (see Figure 7.43). The size of the computational domain is $31Dh \times Dh \times Dh$ (where Dh is the duct hydraulic diameter) for the spatial duct and $6.4Dh \times Dh \times Dh$ for the temporal duct. The corresponding grid-point numbers are, respectively, $318 \times 50 \times 50$ and $64 \times 50 \times 50$. As shown in [251] for the periodic case, the heating significantly influences the topology of the turbulent structures. From Eq. (7.54), the temperature increase induces a global enhancement of the viscous unit, since ρ_w and τ_w are going to decrease. In fact, the near-wall structures such as the low- and high-speed streaks and the associated ejections will keep their size in wall units and hence grow in physical units.[15] The size of the ejections becomes such that they concentrate near the middle plane of the duct. Similar changes are observed in the spatially growing duct. The advantage of this simulation is its ability to visualize the progressive change of the flow structures near the heated wall. The increase in size of the streaky structures is clearly observed.

[15] Notice that in the compressible channel at constant wall temperature just studied, the wall unit did decrease, whereas the wall structures kept their size in physical units.

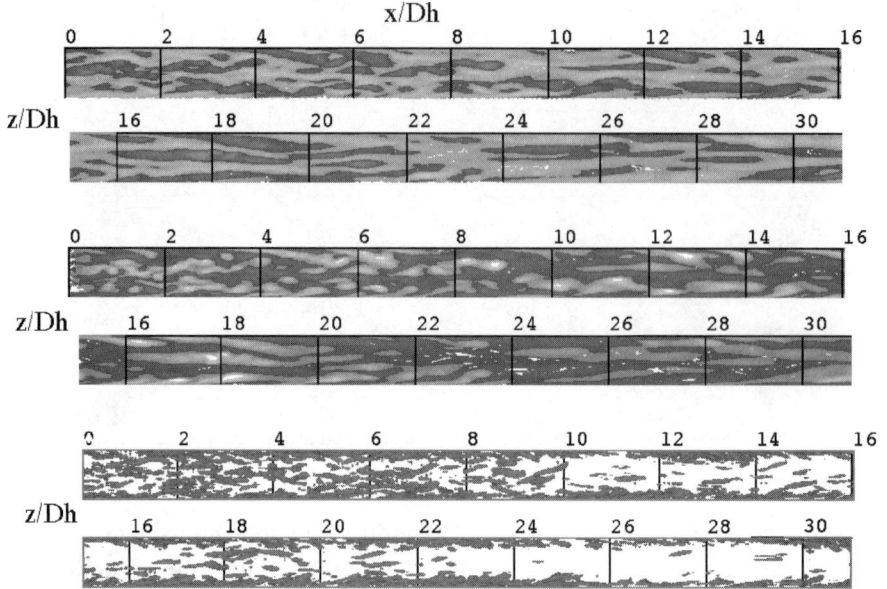

Figure 7.44. Spatially developing heated duct. (Top) fluctuating streamwise velocity near the hot wall; (middle) fluctuating temperature near the hot wall; (bottom) three-dimensional view of isosurface $Q = 0.5(U_b/D_h)^2$ close to the hot wall.

Figure 7.44 shows a view of fluctuating temperature and streamwise velocity in a plane parallel to the hot wall and close to it (at distance $y/D_h = 0.01$) as well as a three-dimensional isosurface of Q in the duct seen from above the heated wall. Positive streamwise velocity fluctuations correspond to sweeps toward the wall and negative fluctuations to ejections. The former transport cold fluid toward the wall, where they bring negative temperature fluctuations; the latter eject hot fluid into the colder outer region, bringing positive temperature fluctuations. Near the inlet, the streaks display their characteristic long and narrow shape. From $x/D_h \approx 6.0$ the streaks are longer and wider than in the inlet region. At the exit the streaks are so wide that only two or three are visible at the end of the duct. Quasi-longitudinal vortices identified with positive Q are very numerous at the inlet, but their number decreases with the streamwise direction as do the velocity and temperature streaks. They concentrate around the middle wall plane, and their longitudinal length is higher close to the outlet. The vortices observed between $x/D_h = 12$ and 30 in the vicinity of the middle wall plane (bottom of Figure 7.44) resemble vortices found in the fully developed heated duct. Figure 7.45 shows the instantaneous temperature field contours and the instantaneous transverse velocity vectors at four different x planes. The evolution of the big ejection is strongly correlated to the increase of the instantaneous secondary flow size. Close to the inlet region, the secondary-flow pattern near the hot wall is similar to the one

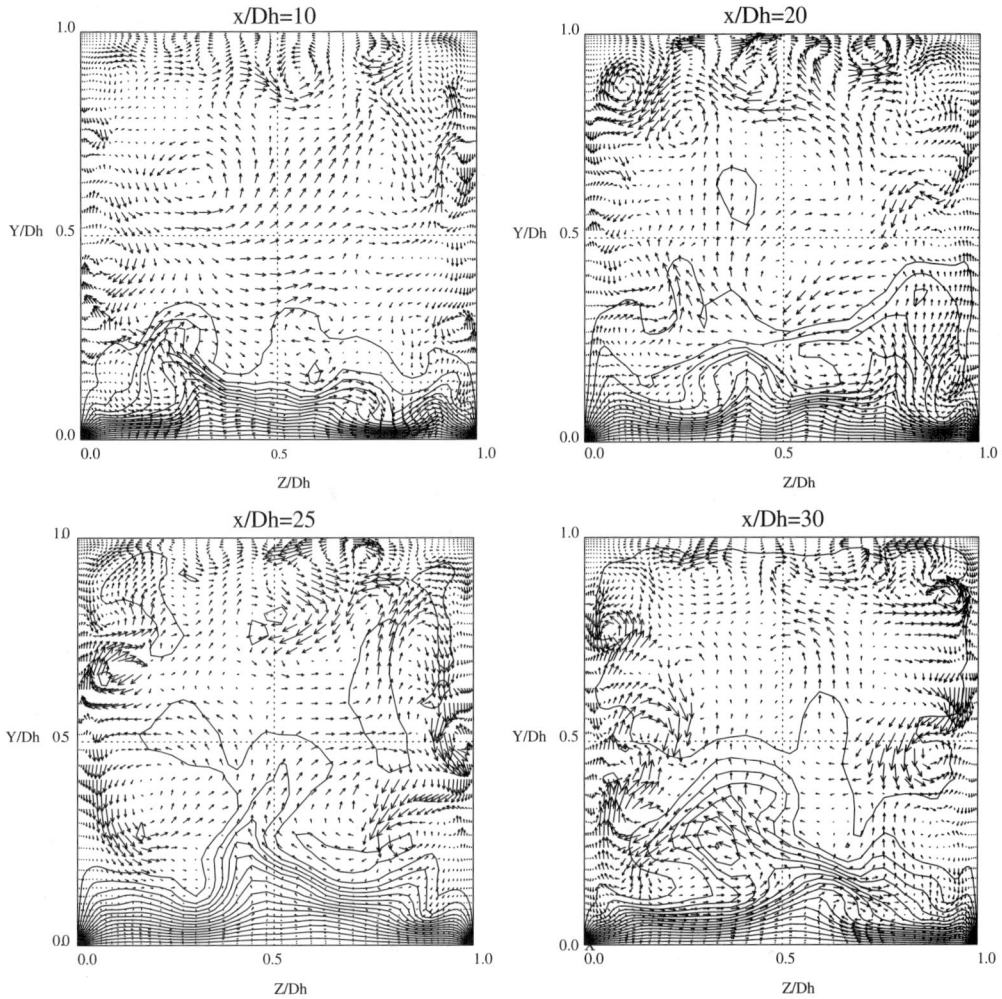

Figure 7.45. Instantaneous temperature and transverse velocity vector fields for $x/D_h = 10$, 20, 25, and 30. The heated wall is located at the bottom of the section. (Courtesy J. Hébrard.)

on the other three walls. Small ejections are observed on the heated wall. However, close to the outlet, a large structure forms that ejects hot fluid from the heated wall.

7.7.4 Curved ducts of square section

Curved ducts are very often encountered in industry, and it is crucial to predict the effects that the curvature may induce on heat exchanges. Various experimental and theoretical studies on turbulent flows in curved ducts have been performed (see, e.g., Hunt and Joubert [128], Hoffman et al. [125], and Saric [256]). It is well known that a concave wall is responsible for Görtler

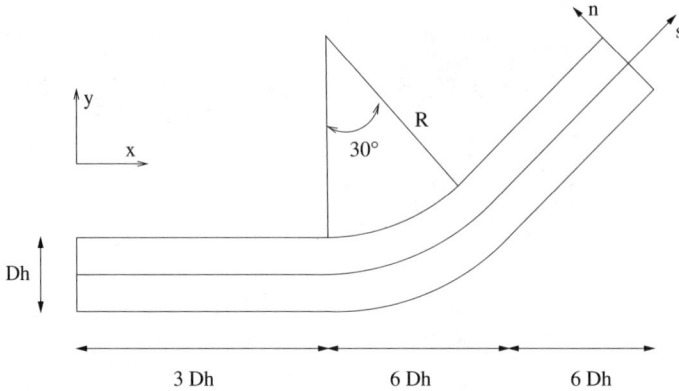

Figure 7.46. Curved duct configuration.

instability, which eventually leads to the formation of Görtler vortices. We have already seen examples of these vortices for *Hermès* rear-flap simulations.[16] However, few numerical works based on LES of a curved duct are available – particularly when the duct is square.

We present LES in a curved duct without heating or when one wall is heated. Details can be found in [214]. The S-shaped duct has also been computed by Hébrard et al. [119]. Here we consider a curved duct of square cross section consisting of a curved part of angle $30°$ and a curvature radius of $10D_h$. The curved part is surrounded by two straight parts: The inflow part has a length of $4D_h$ and the outflow part has a length of $6D_h$ (see Figure 7.46). The Mach and Reynolds numbers are still 0.5 and 6,000, respectively. As was done previously, a fully turbulent flow is obtained at the entry of the duct by injecting the fields issued from a periodic duct computation.

We recall that, in a straight duct of square section, the intensity of the transverse secondary flow reaches a magnitude of about 2% of the mean bulk velocity. In a curved, nonheated duct, the secondary flow is greatly enhanced, reaching 20% of the main flow. As shown in Figure 7.47(b), the destabilization occurs on the concave wall, but the pressure gradient between the inner and outer curved walls implies a displacement of the longitudinal vortices toward the convex side. Two intense Görtler vortices are clearly visible with positive Q isosurfaces. The cross section shown in Figure 7.47(a) reveals the strong amplification of the secondary flow close to the convex wall near the end of the curved part of the duct.

Animation 7-10 gives more details. One sees in particular the system of longitudinal vortices that detaches from the upper convex wall and joins the system of Görtler vortices generated by the lower wall. Animation 7-11 shows the same LES vortices in the vicinity of the concave boundary. Animation 7-12

---

[16] And also for the recirculating flow in a cavity.

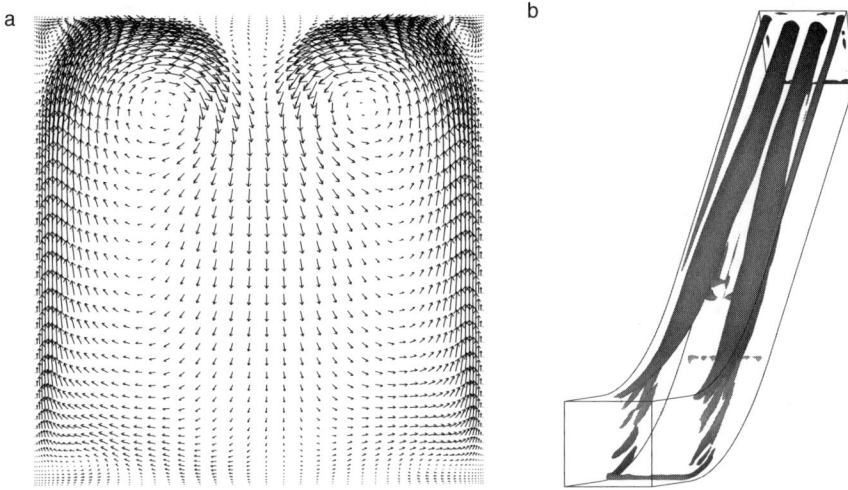

Figure 7.47. Mean results for a nonheated curved channel. (a) Secondary flow after the curvature. (b) Three-dimensional isosurfaces of $Q = 0.3$. (Courtesy C. Münch.)

presents the same view when the concave wall is heated. In this case vortices become bigger and more intense. This is a fine example of the advanced heat-transfer simulations that can be achieved by LES.

7.8 Animations

Animation 7-1: LES of a Mach 0.3 boundary layer spatially developing on a flat plate, showing successively the transition during K-transition (isosurfaces of u', ω'_x, ω'_z) and isosurfaces of u' and low pressure when turbulence has developed. (Film 7-1.mpg; courtesy E. Briand.)

Animation 7-2: The beginning is devoted to the Mach 0.3 boundary layer LES with isosurfaces of u' and Q during, respectively, K- and H-transition. For K-transition, two Q thresholds are compared (0.01 and 0.02). Then u' and Q are shown in the developed region, traveling in a frame moving with velocity $0.6U_\infty$. The rest of the movie concerns models of wave packets. (Film 7-2.mpg; courtesy E. Briand.)

Animation 7-3: LES of the Mach 0.3 channel with presentation of Q isosurfaces and ω_z isolines on the walls. A perspective view of the flow above the flat (top left) and ribbed (bottom right) walls is shown first. Then a view from above is shown. (Film 7-3.mpg; courtesy Y. Dubief.)

Animation 7-4: LES of an obstacle with a wall effect. Isosurfaces of positive Q (green) and positive (red) and negative (blue) longitudinal vorticity are presented. The observer is rotating, and the movie starts with a side view where the flow goes from left to right. (Film 7-4.mpg; courtesy P. Begou.)

Animation 7-5: LES of turbulence in a channel at Mach 0.33 above "high" riblets showing cross section of longitudinal vorticity (red, positive; blue, negative). (Film 7-5.mpg; courtesy G. Hauët.)

Animation 7-6: Same as Animation 7-5 except for a "great" riblet. (Film 7-6.mpg; courtesy G. Hauët.)

Animation 7-7: "High" riblet at Mach 0.33, showing cross sections of positive (yellow) and negative (blue) u' in a plane located 10 wall units away from the riblet peaks. (Film 7-7.mpg; courtesy G. Hauët.)

Animation 7-8: Same as Animation 7-7 except for the "great" riblet. (Film 7-8.mpg; courtesy G. Hauët.)

Animation 7-9: Positive Q isosurfaces in the LES of a transonic flow above a rectangular cavity. Shown are, successively, a perspective view from upstream (with the flow going from left to right), a side view, and a view from downstream. (Film 7-9.mpg; courtesy Y. Dubief.)

Animation 7-10: LES of a curved square isothermal channel at Mach 0.5 showing a global view of the positive Q isosurfaces colored by local longitudinal vorticity. (Film 7-10.mpg; courtesy C. Münch.)

Animation 7-11: LES of a curved square isothermal channel at Mach 0.5 showing a perspective view of the positive Q isosurfaces colored by local longitudinal vorticity close to the concave wall. (Film 7-11.mpg; courtesy C. Münch.)

Animation 7-12: LES of a curved square channel at Mach 0.5 heated on the concave wall showing a perspective view of the positive Q isosurfaces colored by local longitudinal vorticity close to the same boundary. (Film 7-12.mpg; courtesy C. Münch.)

8 Geophysical fluid dynamics

As already stressed, the large-eddy simulation (LES) concept was developed by the meteorologists Smagorinsky, Lilly, and Deardorff. In fact, geophysical and astrophysical fluid dynamics contain an innumerable list of processes (generally three-dimensional) that can be understood experimentally only via laboratory and in situ experiments and numerically mostly by LES. We recall, for instance, that the Taylor-microscale-based Reynolds number for small-scale atmospheric turbulence is larger than 10^4, and thus implementation of DNS does not seem feasible in this case even with the unprecedented development of computers.[1] As far as Earth is concerned, these processes are part of the extraordinarily important issue of climate modeling and prediction involving a very complex system that dynamically and thermodynamically couples the atmosphere (with water vapor, clouds, and hail), the oceans (with salt and plankton), and ice for periods of time from seconds to hundreds of thousands of years. The issue of global warming, which requires our being able to predict the evolution, under the action of greenhouse-effect gases, of Earth temperature (in the average or in certain particular zones), is vital for the survival of populations living close to the oceans and seas. Indeed, global warming induces ice melt,[2] which implies a sea-level elevation. It also increases evaporation, resulting in heavy rains and floods.

We first provide in this chapter a general introduction to geophysical fluid dynamics (GFD). Then we will concentrate on two problems for which LES and DNS provide significant information. The first is the effect of a fixed solid-body rotation on a constant-density free-shear or wall-bounded flow. The second is the generation of storms through baroclinic instability in a dry atmosphere. A third problem that will not be discussed here is oceanic

---

[1] Such simulations would become practical through the use of quantic computers with binary information carried out by atoms.

[2] Ice melt has been observed everywhere from glaciers over many years to polar ice fields.

deep-water formation, the essential link of the oceanic conveyor belt in the northern Atlantic, where the Gulf Stream water becomes saltier[3] and sinks by gravity. Large-eddy simulations related to the latter problem may be found in Padilla-Barbosa and Métais [228].

8.1 Introduction to geophysical fluid dynamics

8.1.1 Rossby number

We first consider a flow that rotates with a constant angular velocity $\Omega = f/2$. We work in the rotating frame. Let U be a characteristic relative velocity of the fluid and let L be a characteristic scale of motion. We define the Rossby number as the ratio of characteristic inertial over Coriolis accelerations in the Navier–Stokes equations, which yields

$$Ro = \frac{U}{fL}. \tag{8.1}$$

When $Ro \gg 1$, rotation is negligible. When $Ro \ll 1$, rotation dominates. In a rotating flow on a sphere at a latitude φ, the flow is assumed to be approximately equivalent to a flow rotating around the local vertical axis with an angular velocity $\Omega \sin \varphi$ defined by the projection of the rotation vector $\vec{\Omega}$ on the local vertical. Then the Rossby number is still defined by Eq. (8.1) with $f = 2\Omega \sin \varphi$ being the Coriolis parameter.

Here, we will take horizontal quantities for U and L.

8.1.2 Earth atmosphere

Large atmospheric scales
The large scales are called synoptic with horizontal wavelengths of the order of, or larger than, several hundred kilometers. Because the effective thickness of the atmosphere is of the order of 15 km,[4] these motions are on a shallow layer and are quasi-two-dimensional on the Earth's sphere. In medium latitudes, synoptic motions correspond to quasi-horizontal vortices rotating around zones of high or low pressure, respectively, in the anticyclonic or cyclonic sense.[5] This circulation is driven by the geostrophic balance between the pressure gradient and the Coriolis force in the motion equations. Cyclonic vortices are more energetic than anticyclonic ones. The associated Rossby

---

[3] The water becomes saltier because a part of it is transformed into ice, which is fresh.
[4] This height corresponds to about 80% of the atmospheric mass. Above this level air is more and more rarefied.
[5] Cyclonic means here the same sense of rotation as that of Earth: anticlockwise in the Northern Hemisphere and clockwise in the Southern Hemisphere.

number ($U = 30$ m/s, $L = 1,000$ km, $f = 10^{-4}$) is 0.3. The origin of these vortices is the baroclinic instability of easterly jet streams. These are encountered by planes while flying at a 10-km elevation. The jet streams arise from the so-called thermal-wind equation. It results from combining the horizontal geostrophic balance and the vertical hydrostatic balance between the pressure gradient and gravity (see Lesieur [170] for more details). The thermal-wind intensity is proportional to the horizontal north–south temperature gradient. Its regular intensity is 30 m/s, but it may reach much higher values, such as the 110 m/s recorded in the days before the great storms of December 26 and 28, 1999, in Europe.

Tropical cyclones

At lower latitudes, tropical cyclones also correspond to the production of cyclonic vortices, but the Rossby numbers are larger than in the former case. The main effect here is the substantial evaporation of water at the sea surface if its temperature is greater than 26 °C. The water vapor thus produced rises owing to thermal convection and condenses higher up because of lower temperatures with the latent heat released being converted into horizontal kinetic energy, which drives the system. With $U = 60$ m/s, $L = 200$ km, and $f \approx 10^{-4} \sin 23° / \sin 45°$, one gets $Ro \approx 5$.

Hadley cells and trade winds

The strong thermal convection above the ocean in the intertropical zone is responsible for the formation of two cells on both sides of the equator called Hadley cells: Warm air rises in the equatorial region, travels to higher latitudes where it cools, and then descends at the Tropics. While descending, it deviates westward because of the Coriolis force, giving rise to trade winds. Weaker cells rotating in the opposite sense, called Ferrel cells, are also observed. This is in fact a year–time-averaged view. Seasonal variations of Hadley cells lead to the monsoon phenomenon, where trade winds cross the equator and change sign, owing to the opposite direction of the Coriolis force. When these phenomena occur in a continent bordered to the west by an ocean, the monsoon is accompanied by heavy rains.

Ozone hole

Let us mention the existence of Arctic and Antarctic circumpolar vortices. These are cyclonic. Like the jet streams, they obey the thermal-wind equation. The north–south temperature gradient is very important at the poles because of the existence of ice. Therefore the Antarctic circumpolar vortex is particularly intense. It is within this vortex that the seasonal phenomenon known as the "ozone hole" occurs, which is generally explained by the action of chlorofluorocarbon (CFC) gases released at medium latitudes. These CFCs rise in the atmosphere to the stratosphere (which is very stable from a thermal point of view with a permanent inversion) and then travel by quasi-horizontal turbulent diffusion everywhere – particularly to the southern pole. For reasons still unclear they can cross the border of the vortex (although it is very well marked) and penetrate inside. Ozone destruction occurs at the beginning of austral spring (end of September) through complex mechanisms that are far from being understood. This phenomenon is particularly marked since 1980. Because ozone in the stratosphere protects us from ultraviolet radiation coming from the sun, drastic international measures have been taken to forbid the production and use of CFC. However, the southern ozone hole does not show any tendency to disappear. There is also now a weaker Arctic ozone hole during spring.

Mesoscale and small-scale meteorology

This scale involves wavelengths ranging from 10^{-3} m (the Kolmogorov scale) to several tenths of kilometers. These motions are strongly three-dimensional. They are affected by thermal stratification and sometimes rotation. Thermal stratification may be stable in inversion zones or convectively unstable (in thunderstorms or tornadoes[6]). Durable inversion layers are often observed above cities located in troughs or surrounded by mountains, such as Grenoble, Los Angeles, or Mexico City, and are responsible for substantial industrial and car pollution. In this context, an interesting unstationary RANS has been carried out by Kenjeres and Hanjalic [140]. Let us mention also the LES of Fallon et al. [90] using the SSF model of a stably stratified flow passing a straight backstep.

In an inversion situation, wind crossing a mountain is going to give rise to internal gravity waves called lee waves, the breakup of which is an important source of turbulence. Because these waves propagate upward, they are responsible for the so-called clear-air turbulence met by planes while passing above mountains at elevations of the order of 10 km. Finally, the atmosphere in contact with the ground gives rise to a turbulent boundary layer affected by rotation. It is a turbulent Ekman layer, whose typical height is 1 km; again, it may be strongly felt during the landing of a plane.

8.1.3 Oceanic circulation

Large-scale circulation

Oceanic circulation at planetary scales is forced by winds. Trade winds, in particular, entrain surface water westward. In the northern Atlantic for instance,

---

[6] Let us evaluate the Rossby number of a developed tornado at medium latitude: Taking $U = 60$ m/s and $L = 1$ km, we have $Ro = 600$.

recirculation of the north-equatorial current within the Gulf of Mexico creates the Gulf Stream. Because the dominant winds at higher latitudes and close to the ocean are mainly westerly,[7] this yields the formation of a large anticyclonic recirculation cell characterized by a warm current rising on the western border of the basin, then crossing it up to the eastern border, and then going down to the equator. In the northern Pacific, the equivalent of the Gulf Stream is the Kuroshivo, and the current traveling to the equator is the California current. The latter is characterized by upwellings of cold, deep water, as the Coriolis force diverts the warm surface water entrained by the wind to the open sea; this water, then, by continuity, has to be replaced by deep water. Upwelling regions are also, because of the strong horizontal temperature gradients involved, subjected to baroclinic instability with production of oceanic vortices. Let us mention in this respect the LES of Tseng and Ferziger [282], who look at the effect of coastal shape on vortices generated by upwellings. The equivalent of the California current in the southern Pacific is the cold Humboldt current traveling to the equator along the coast of Peru. Cold upwelling currents are extremely rich in fish because they contain more oxygen. Upwelling exists also off the western coast of France, Portugal, and Africa, but the most famous of all is the Humboldt current, whose anomaly called El Niño is characterized by a reversal of the current, which becomes warm, causing fish to disappear with deleterious consequences for affected national economies. The quasi-period of this unpredictable event is between 2 and 4 years. El Niño seems to be associated with a sort of nonlinear oscillation of the coupled atmosphere–ocean system and has been observed for many centuries. The problem is that it seems to become more and more intense – perhaps as a consequence of global warming.

A few words on the aforementioned oceanic conveyor belt are in order. Currently, an important warming of the northen Atlantic ocean is observed with the threat of significant ice melt. This may reduce the deep-water formation in such a way that the Gulf Stream would no longer descend. However, we recall that the engine of the oceanic system is made of trade winds, which rather than being reduced by global warming, would be amplified because they result from thermal convection. So the Gulf Stream would certainly not disappear, but it might, through continuity, have to find other routes than the present one.

Mesoscale and small-scale oceanography

Oceanic currents are subjected to various instabilities responsible for the formation of eddies of scale 50–100 km. Taking $U = 10$ cm/s and $L = 50$ km, we find that oceanic vortices at medium latitudes have a Rossby number of 0.02 – about ten times smaller than in the synoptic atmosphere. This is of the

[7] The winds are westerly because of both Ferrel cells and jet streams.

same order as Jupiter's Great Red Spot: Here one takes $U = 100$ m/s, $L = 20,000$ km, and $\varphi = 45°$. Jupiter rotates faster than Earth with an approximate period of 10 h. We have $f = 10^{-4} \times 24/10$, which gives $Ro \approx 0.02$, of the same order as oceanic vortices (see Somméria [272]).

Because the error in making the geostrophic-balance assumption is proportional to the Rossby number, this calculation shows that Earth's oceanic vortices or Jupiter's vortices are closer to geostrophic balance than to a synoptic atmosphere.

At smaller scales, the ocean may be the seat of intense three-dimensional turbulence, resulting, for instance, from the breakup of internal-gravity waves on the coast. Turbulent Ekman layers exist also close to the bottom and at the surface where the wind blows.

8.1.4 Internal geophysics

Turbulence exists in the strongly heated liquid metal of Earth's outer core. Here, the Rossby number may be calculated as follows at medium latitudes (Cardin [36]). One takes $U = 10^{-3}$ m/s, and $L = 1,000$ km, which gives $Ro = 10^{-5}$. These flows are, because of their extreme slowness, the most rotation dominated of all those already considered up to now. They are electrically conductive and obey magnetohydrodynamic (MHD) equations. Furthermore they are subjected to a strong internal convection, resulting in quasi-two-dimensional vortices of axis parallel to Earth's axis of rotation.

8.2 Effects of spanwise rotation on shear flows of constant density

We study now with the aid of DNS or LES shear flows of uniform density rotating about a spanwise axis. The flow is assumed periodic in the streamwise x and spanwise z directions with $\bar{u}(y, t)$ being the longitudinal velocity at y averaged in the streamwise and spanwise directions. Here v and w are the velocity components along y and z, respectively. We define now a local vorticity-based Rossby number

$$Ro(y, t) = -\frac{1}{2\Omega}\frac{d\bar{u}}{dy} \qquad (8.2)$$

as the ratio of the spanwise relative vorticity to the entrainment vorticity. Regions with a positive (or negative) local Rossby number will be called cyclonic (or anticyclonic). Recall that the absolute vorticity vector $\vec{\omega}_a = \vec{\omega} + 2\Omega\vec{z}$ satisfies Helmholtz's theorem in its conditions of applicability, which stresses that absolute-vortex elements follow the fluid parcels they contain. The Rossby number $Ro^{(i)}$ is the minimal value of $Ro(y, 0)$.

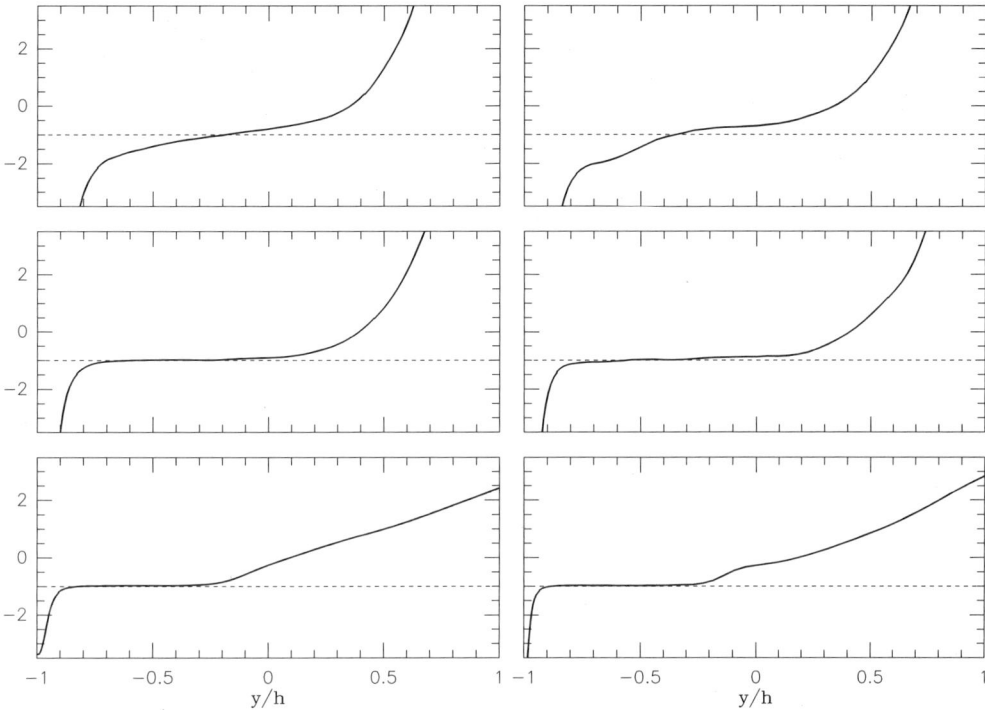

Figure 8.1. Rotating channel of uniform density. Eventual local Rossby number distribution in the DNS (left) and LES (right). From top to bottom, minimal Rossby numbers of -18, -6, and -2. (From Lamballais et al. [154].)

A pioneering linear-stability analysis of the problem in the inviscid case and for longitudinal modes (x-independent) was carried out by Pedley [233]. Details are given in Lesieur ([170], p. 73) with analogies to centrifugal instabilities. Pedley [233] shows that a necessary and sufficient condition for instability is that $Ro(y, 0) < -1$ somewhere in the flow.

8.2.1 Rotating channel

As in the nonrotating uniform-density case, we still consider constant-flow-rate DNS or LES for which we take as initial conditions a randomly perturbed parabolic profile. The axis $y = 0$ is the channel centerline. Thus $Ro(y, 0)$ is linear and antisymmetric with respect to y. We choose the direction of $\vec{\Omega}$ such that the region of the channel $y > 0$ is initially cyclonic; $y < 0$ is anticyclonic. The minimum value $Ro^{(i)} = Ro(-h, 0)$ is always negative. Lezius and Johnston [180] have shown that such a flow is inviscidly unstable if $Ro^{(i)} < -1$. This is in fact equivalent in this case to Pedley's [233] result.

Figure 8.1, taken from Lamballais et al. [154], shows the distribution of the local Rossby number in DNS and LES as a function of $Ro^{(i)}$. The DNS is carried out at a Reynolds number based on the bulk velocity and $2h$ of 5,000.

The LES is done for the same conditions as in the nonrotating LES using the spectral-dynamic model presented in Chapter 5 at a Reynolds number of 14,000. We see that a $Ro(y) = -1$ plateau forms for the three rotation rates in the LES case and for the last two for the DNS, yielding

$$\frac{d\bar{u}}{dy} = 2\Omega, \tag{8.3}$$

which corresponds to a zero spanwise absolute vorticity. This result was shown experimentally by Johnston et al. [137]. Numerically, one should quote the DNS studies of Kim [141], Tafti and Vanka [277], Kristoffersen and Andersson [148], Nakabayashi and Kitoh [217], and the LES of Piomelli and Liu [237] using Smagorinsky's dynamic model. But the simulations of Lamballais et al. [154] investigate lower Rossby number moduli (which means faster rotation rates) than do these authors. The DNS results of Lamballais are presented in Lesieur ([170], pp. 432–433) with pictures of the vorticity modulus compared with the nonrotating case and also of the mean velocity profiles. Lesieur notes: "It is clear that the flow is quasi-laminar on the cyclonic side, while hairpins on the anticyclonic side are more and more inclined with respect to the wall as rotation is increased. It was also checked that longitudinal velocity fluctuations on this side are reduced when the Rossby number is increased, and that the corresponding streaks have disappeared at $Ro^{(i)} = -2$."

This order of magnitude of Rossby number (< -1) is what we call here "moderate" rotation. As stressed by Lezius and Johnston [180], there is a lower initial Rossby number (inferior to -1) under which the flow is stable again and that decreases as the Reynolds number is augmented.

If one further increases the rotation rate,[8] there will be an interesting crossover at $Ro^{(i)} = -1$. Indeed, the case $Ro^{(i)} \geq -1$ ("fast" rotation) is totally different because it is stable and two-dimensionalizing from the point of view of the aforementioned instability. However, Lamballais et al. [152] have shown it may be subject to the growth of two-dimensional TS waves, as is the case in particular for a DNS at $Ro^{(i)} = -0.1$: TS waves reach a nonlinear saturated state, and the flow is composed of purely two-dimensional spanwise vortices of alternate-sign vorticity on the sides of the channel. The two rows of vortices on each wall are out of phase. This is analogous to a purely two-dimensional solution studied by Jimenez [134].

Finally, we remind the reader that, for the channel, $Ro^{(i)}$ cannot exceed zero by definition.

---

[8] For the channel, $Ro^{(i)}$ is always negative, and thus increasing it corresponds to an increase in the rotation rate.

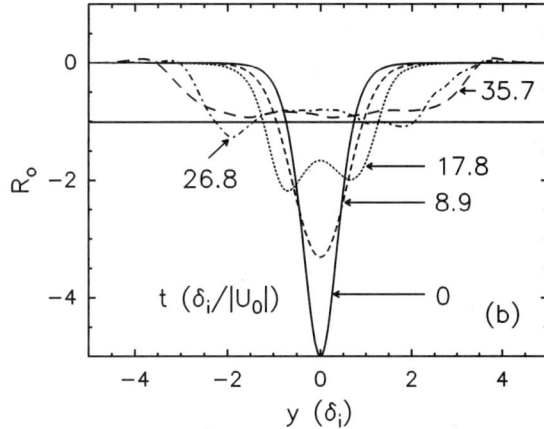

Figure 8.2. Anticyclonic rotating mixing layer showing time evolution of the local Rossby number in the DNS of Métais et al. [206]. (Courtesy *J. Fluid Mech.*)

8.2.2 Rotating free-shear layers

Strong analogies exist with mixing layers and wakes rotating about a spanwise axis, where Pedley's analysis is also valid in the inviscid case. This was complemented by the viscous linear-instability studies of Yanase et al. [294]. They show that the KH instability is suppressed and replaced by the "shear-Coriolis" instability, a purely longitudinal instability, if $Ro^{(i)}$ is, again, strictly lower than -1. For the channel, there is a minimum lower bound for $Ro^{(i)}$ in this instability, which decreases as the Reynolds number increases. Isoamplification rates for the mixing layer taken from this study are given in Lesieur ([170], p. 424).

Rotating mixing layer

For $Ro^{(i)} < -1$, and if the Reynolds number is high enough, the shear-Coriolis instability manifests itself. The DNS and LES of Flores [96] and Métais et al. [206] show then that in these conditions in the anticyclonic regions there is a stretching of intense purely longitudinal alternate vortices of absolute vorticity, which are such that the mean spanwise absolute vorticity becomes zero and the Rossby number becomes equal to -1 over a large fraction of the region (see Lesieur [170], pp. 430–431). This is clear from Figure 8.2, which shows the time evolution of the local Rossby number in a rotating mixing-layer DNS carried out by Métais et al. [206]. The Rossby number peaks initially, following the initial vorticity distribution. Then the amplitude of the peak decreases while the latter widens. At $t = 26.8$, a plateau close to -1 (slightly higher in fact) forms; this plateau is still there at $t = 35.7$.

Now we increase $Ro^{(i)}$. There is again a crossover at $Ro^{(i)} = -1$, above which the KH instability occurs, with a strong two-dimensionalization: Helical pairing is suppressed, as well as hairpin stretching between KH vortices.

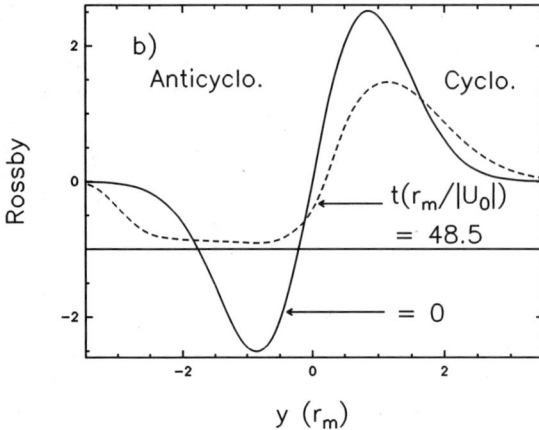

Figure 8.3. Rotating-wake DNS at $Ro^{(i)} = -2.5$ showing local Rossby number profiles at $t = 0$ and $t = 48.5$.

The structure of the mixing layer is purely two-dimensional. Here, and in contrast to the channel, $Ro^{(i)}$ may become positive. In this case the mixing layer has become cyclonic, and increasing $Ro^{(i)}$ corresponds now to a reduction of the rotation rate. Our DNS and LES (see also Lesieur et al. [166]) do show that it remains very two dimensional up to $Ro^{(i)}$ of the order of 10. It is clear that if the Rossby number is increased and goes to infinity, three-dimensional instabilities of the type found in the nonrotating case will develop again.

Rotating wake

The DNS and LES of Flores [96] and Métais et al. [206], as well as unpublished calculations done in Grenoble, enable us to stress the following conclusions for the wake: For $Ro^{(i)} < -1$, as in the mixing layer, and if the Reynolds number is high enough, the shear-Coriolis instability takes hold. Because the wake has both cyclonic and anticyclonic sides, the Karman street is deeply modified. On the cyclonic side, one observes a two-dimensional row of vortices without secondary hairpin-vortex stretching. On the anticyclonic side, however, the anticyclonic Karman vortices no longer exist. They are replaced by the same longitudinal vortices as for the anticyclonic mixing layer at $Ro^{(i)} < -1$ discussed earlier, with the local Rossby number becoming equal to -1 in this range. This is clear from Figure 8.3, taken from a rotating-wake DNS at $Ro^{(i)} = -2.5$ and showing at times $t = 0$ and $t = 48.5$ the local Rossby number profiles. Here U_0 is the initial maximum deficit velocity and r_m is a typical initial wake width. One sees how the initial antisymmetric Gaussian Rossby distribution is transformed, with on the anticyclonic side, a plateau slightly higher than -1, as in the anticyclonic mixing layer.

The crossover at $Ro^{(i)} = -1$ exists again. Above, this, the wake becomes a purely two-dimensional Karman street, and there is no great difference in the wake structure between $Ro^{(i)} = -1$ and $Ro^{(i)} = -0.1$. As for the channel, $Ro^{(i)}$ cannot exceed zero.

Universality of free-shear layers

These results demonstrate a very interesting universality of free or wall-bounded shear layers rotating about a spanwise axis as far as the three following points are concerned:

- There is a crossover Rossby number $Ro^{(i)} = -1$ separating a structure dominated by Pedley's [233] longitudinal mode if the Reynolds number is high enough from a two-dimensional structure.
- Under the crossover, there is a region of space where one observes the establishment of a $Ro(y) = -1$ plateau. This point, well predicted by LES and DNS, still poses severe problems to one-point closure models (see, e.g., Nagano and Hattori [216]).
- In this region, the flow evolves into a set of purely longitudinal absolute vortices.

This last point has been demonstrated numerically. The explanation of the second point provided by many authors for the channel case[9] is based on the fact that turbulence evolves toward a marginally stable state from the point of view of linear-stability theory. Another explanation is given by Lesieur et al. [173] in terms of nonlinear reorientation of absolute vortices. They propose an exact analysis based on Euler equations, where x-independence[10] is assumed. The evolution equations (following the motion) of the absolute vorticity $\vec{\omega}_a$ of components $\omega_1 = \partial w/\partial y - \partial v/\partial z$, $\omega_2 = \partial u/\partial z$, and $\omega_3 + f = -\partial u/\partial y + f$ can be written for this x-independent solution as

$$\frac{D}{Dt}\vec{\omega}_a = \overline{\overline{F}} \otimes \vec{\omega}_a \tag{8.4}$$

with

$$\overline{\overline{F}} = \begin{pmatrix} 0 & f & 0 \\ 0 & \dfrac{\partial v}{\partial y} & \dfrac{\partial v}{\partial z} \\ 0 & \dfrac{\partial w}{\partial y} & \dfrac{\partial w}{\partial z} \end{pmatrix} = \overline{\overline{F}}_1 + \overline{\overline{F}}_2, \tag{8.5}$$

---

[9] Results obtained in Grenoble for the rotating free-shear flows are new and have not been commented on by other people.

[10] Indeed, we have seen that the linear-stability analysis shows that this longitudinal mode dominates shear instabilities in this case.

$$\overline{\overline{F}}_1 = \begin{pmatrix} 0 & f & 0 \\ 0 & 0 & 0 \\ 0 & 0 & 0 \end{pmatrix}, \tag{8.6}$$

and

$$\overline{\overline{F}}_2 = \begin{pmatrix} 0 & 0 & 0 \\ 0 & \dfrac{\partial v}{\partial y} & \dfrac{\partial v}{\partial z} \\ 0 & \dfrac{\partial w}{\partial y} & \dfrac{\partial w}{\partial z} \end{pmatrix}. \tag{8.7}$$

The action of $\overline{\overline{F}}_1$ upon $\vec{\omega}_a$ is to leave its projection $\vec{\omega}_n$ on the yz plane unchanged and to stretch ω_1 as

$$\frac{D\omega_1}{Dt} = f\omega_2. \tag{8.8}$$

We have also

$$\frac{D}{Dt}\vec{\omega}_n = \overline{\overline{F}}_2 \otimes \vec{\omega}_n. \tag{8.9}$$

The tensor $\overline{\overline{F}}_2$ is in fact the velocity-gradient tensor in the yz plane, and we can apply the same analysis as in Chapter 2. Indeed, during the linear stage of evolution, the DNS studies by Métais et al. [206] studies of an anticyclonic mixing layer of initial Rossby number of -5 show the growth of the longitudinal mode with absolute vortex filaments in phase and inclined approximately 45° above the horizontal plane. This produces concentrations of longitudinal vorticity in the yz plane. Let us assume that a nonlinear regime is reached at which longitudinal vorticity concentrations are strong enough to form vortices, whose core is "elliptic" in the sense that the eigenvalues of $\overline{\overline{F}}_2$ (or $-\overline{\overline{F}}_2|^t$) are purely imaginary. Rotation of $\vec{\omega}_n$ about $\vec{x}$ (in the sense of the sign of the longitudinal vorticity) will therefore dominate deformation in Eq. (8.9), implying an increase of the spanwise absolute-vorticity component (which is negative). The Rossby number (which was lower than -1) will increase also. We have here for the absolute-vorticity vector an interesting mechanism of longitudinal self-reorientation possible only in a nonlinear regime.

GFD applications

These results have important applications in geophysical and astrophysical fluid dynamics when density differences may be neglected.

In the mesoscale atmosphere of Earth, typical Rossby number moduli are larger than one. Then anticyclonic vortices should be destroyed and cyclonic ones should be two-dimensionalized. This is observed in the wake (visualized by clouds) of some islands, which display very asymmetric Karman streets.

In the ocean, cyclonic and anticyclonic mesoscale eddies, which do not result from baroclinic instability, should be two-dimensionalized. This might be the case in particular for detached vortices behind capes. An example is the Strait of Gibraltar separating the Atlantic Ocean from the Mediterranean Sea. The Atlantic is much fresher than the Mediterranean.[11] Therefore, Mediterranean water will descend into the Atlantic through Gibraltar, whereas Atlantic water will enter the Mediterranean, remaining at the surface. The coast of Morocco will act as a backward-facing step, and vortices will be shed on the north coast of Algeria. These vortices have been observed. They are anticyclonic, but since the associated Rossby numbers are very low they should be two-dimensionalized.

On Jupiter, where there is no evidence of baroclinic instability, vortices should be two-dimensionalized regardless of their sign. This is the case in particular of the Great Red Spot, which is anticyclonic.

8.3 Storm formation

As already stressed, storm formation results from baroclinic instability of a jet stream in thermal-wind balance resulting from a horizontal north–south temperature gradient. The simplest model for such a study is the Eady model.

Notice that we are going to change the notation with respect to the preceding section: x, y, and z will be, respectively, the zonal, meridional, and vertical directions, and u, v, and w are the velocity components in these directions.

8.3.1 Eady model

Details of the Eady model are given in Drazin and Reid [74]. One considers a channel rotating about the vertical axis z with an angular velocity $f/2$ studied within the Boussinesq approximation. The flow is stably stratified along the vertical and the horizontal with constant potential temperature gradients. The channel has a width L and a depth H. Periodicity is assumed in the x direction. Free-slip boundary conditions on the lateral walls and upper lid are taken with a no-slip condition at the ground. One assumes initially a thermal-wind

[11] Indeed, the latter is very warm and undergoes strong evaporation.

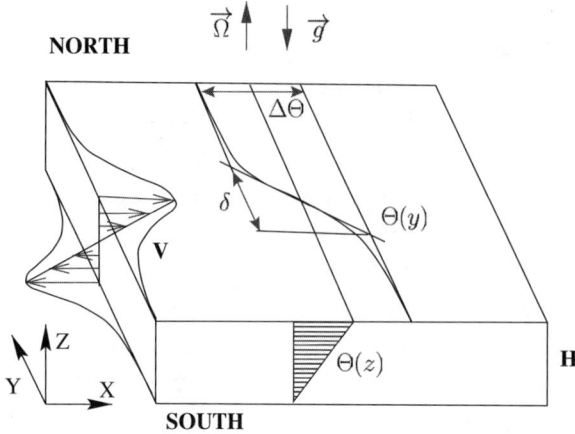

Figure 8.4. Baroclinic jet configuration. (Courtesy E. Garnier.)

balance corresponding to an (x, y)-independent zonal westerly velocity rising linearly from zero at the ground to U on the lid. Let $Ro = U/(fL)$ be the Rossby number, and let $F = U/(NH)$ be the Froude number, where N is the Brunt–Väisälä frequency (see details in Lesieur [170], p. 49). A linear-stability analysis shows that instability occurs when $Ro/F < 0.76$.

8.3.2 Baroclinic jet

The configuration of this simulation (DNS or LES) is displayed in Figure 8.4. It is a baroclinic jet in a channel. One still works within the Boussinesq approximation.

As in the Eady model, there is initially a positive constant vertical temperature gradient. The difference from the Eady model is the presence of a hyperbolic-tangent velocity profile of width δ in the meridional direction.[12] The initial velocity field is in thermal-wind balance and corresponds to a zonal jet varying linearly with the vertical from $-V$ at the ground ($z = 0$) to $+V$ on the lid ($z = H$). We stress that a Galilean transformation of velocity V would yield a zero velocity at the ground, as in reality, and a velocity $2V$ at the lid. The Rossby and Froude numbers are, respectively, $V/f\delta$ and V/NH. The DNS carried out by Garnier et al. [106] allowed (by cancellation of the nonlinear terms) a linear-stability analysis to be made. It shows that instability occurs for $Ro/F < 1.5$. Afterward, simulations are carried out for $Ro/F = 0.5$ corresponding to a physically realistic situation in Earth's atmosphere. The DNS of Figure 8.5 shows the formation of quasi-two-dimensional cyclonic vortices and weaker anticyclonic vortices. The figure also displays the production of intense cyclonic vertical vorticity within the fronts separating the cold fluid to the north from the warm fluid to the south at the ground and under the

---

[12] This width is defined exactly in the same way as for the vorticity thickness in a mixing layer.

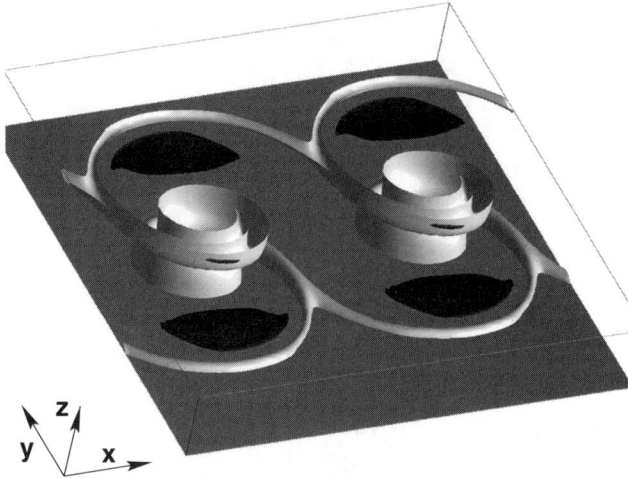

Figure 8.5. DNS of the baroclinic jet. Light gray, cyclonic relative vertical vorticity; black, anticyclonic vorticity. (Courtesy E. Garnier.)

lid. Such a production may be explained as follows: One can show within the framework of Boussinesq approximation that relative vertical vorticity ω satisfies

$$\frac{D\omega}{Dt} = (\omega + f)\frac{\partial w}{\partial z} + \omega_1 \frac{\partial w}{\partial x} + \omega_2 \frac{\partial w}{\partial y}, \qquad (8.10)$$

where ω_1 and ω_2 are here the components of $\vec{\omega}$ along x and y, respectively. It was shown by Garnier et al. [106] in their simulations that the last two terms on the r.h.s. of Eq. (8.10) can be neglected, and the equation reduces to

$$\frac{D}{Dt}\omega \approx (f + \omega)\frac{\partial w}{\partial z}. \qquad (8.11)$$

One understands thus how warm fluid at the ground in contact with the cold fluid will be obliged to rise, yielding $\partial w/\partial z > 0$ on the r.h.s. of Eq. (8.11). Here f is positive, and the Rossby-number modulus of the calculation ($\approx |\omega|/f$) is low enough that $f + \omega > 0$ and ω will grow. Starting with a weak $|\omega|$, there will be growth of cyclonic vorticity and a damping of the anticyclonic-vorticity modulus. The same thing occurs under the lid, where cold fluid will sink under warm fluid with again $\partial w/\partial z > 0$.

This is illustrated by Animation 8-1, which is another DNS done by Garnier at Rossby and Froude numbers both equal to 0.1. It presents the vertical relative vorticity (pink positive, cyclonic; blue negative, anticyclonic). First, we can see the double Bickley jet at the ground and on the lid. Afterward the progressive formation of quasi-two-dimensional cyclonic vortices is observed with formation of intense cyclonic vorticity braids along the fronts.

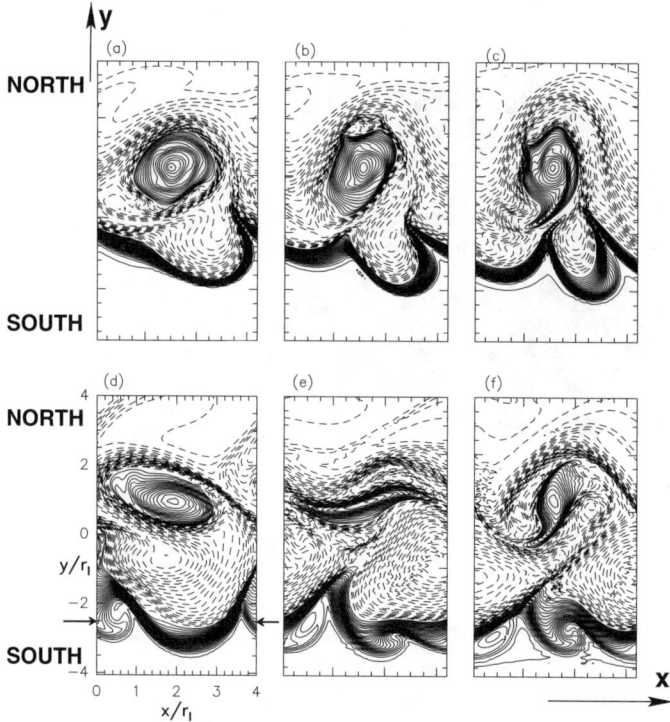

Figure 8.6. Time evolution in two days of the ground temperature in the baroclinic jet LES. Time increase as indicated by the letters. (Courtesy E. Garnier.)

Garnier et al. [106] carried out LES studies using the subgrid model combining the SF model and a hyperviscosity. A primary instability occurs as in the DNS. Then there is a secondary instability of the cold front, with production of two secondary vortices in two days (see Figure 8.6). Their vorticity is of the order of 2 in 3 times that of the primary vortices. Garnier et al. also show that the secondary instability is associated with regions where the ratio of local Rossby to local over Froude numbers is smaller than 1.5, the critical value from the point of view of their linear-stability study.

As stressed in Lesieur et al. [175], there are strong analogies with storms that struck Europe on December 26 and 28, 1999 (see Figures 8.7 and 8.8).

The situation on December 25 (Figure 8.7) indicates a large cyclonic perturbation arriving over Scandinavia after having crossed Great Britain. The corresponding winds, of the order of 30 m/s, are typical of "regular" storms issuing from a primary baroclinic instability that often cross the channel sea. It is clear that France is split by the cold front at Brittany. On Christmas evening in 1999 a warm wind coming from the south was blowing above Grenoble, with temperatures of the order of 15 °C, which is very unusual for the season. This wind was induced by the vortex sitting above Scandinavia. The December 26 storm (Figure 8.8) is a small vortex of ≈400-km diameter, which formed

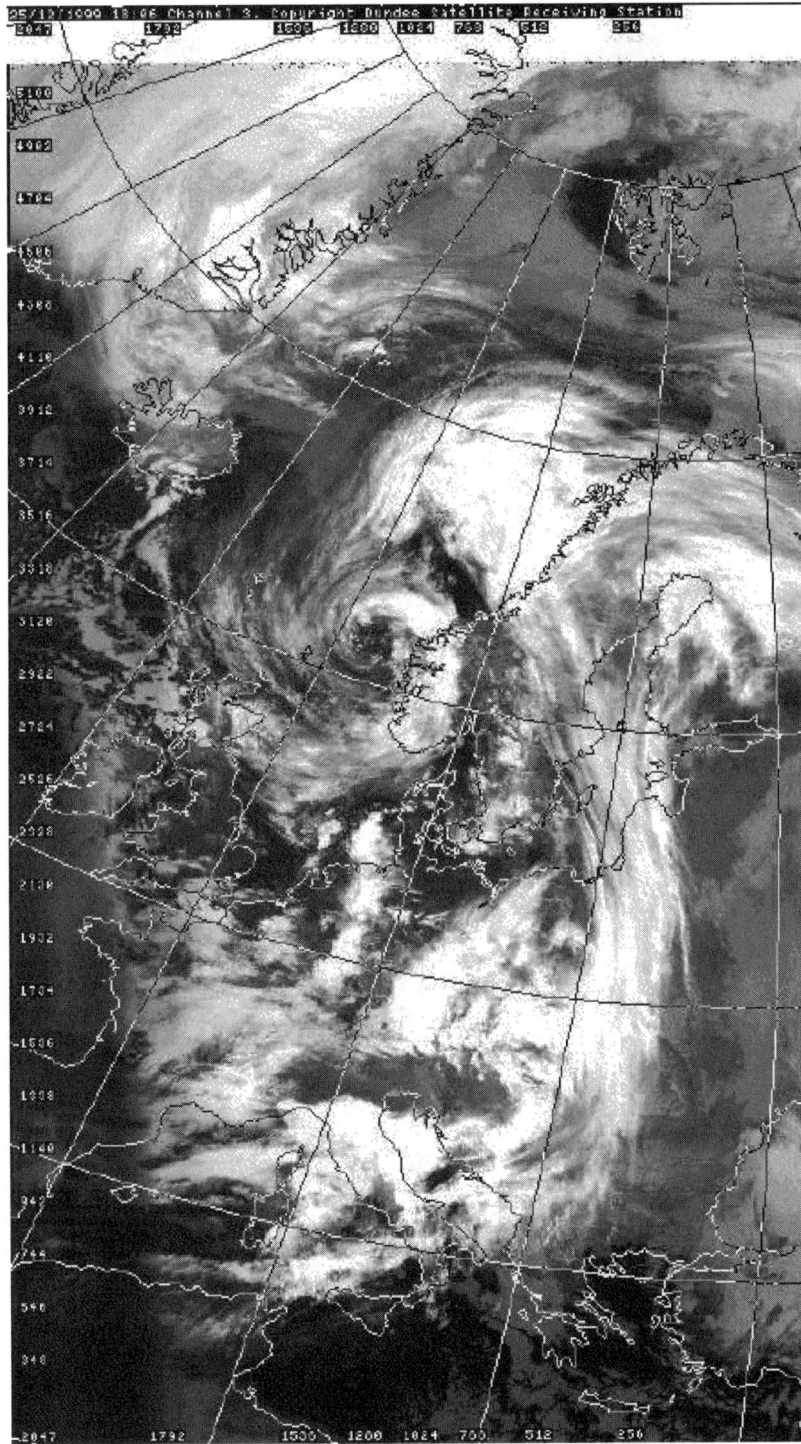

Figure 8.7. Satellite view of Europe on December 25 at 18h. (Courtesy Dundee University.)

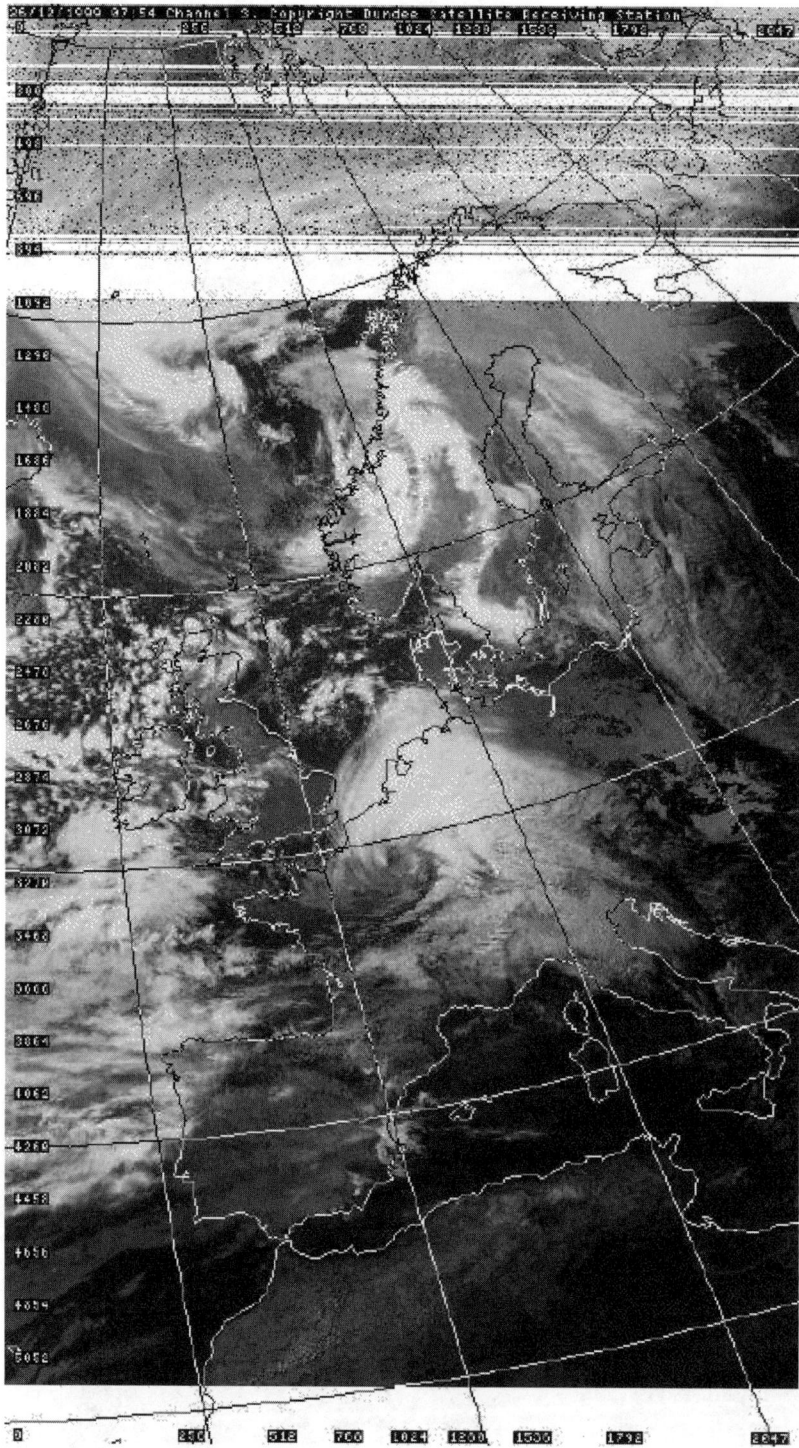

Figure 8.8. Satellite view of Europe on December 26 at 8h. (Courtesy Dundee University.)

above Brittany and, according to Météo-France, crossed France from west to east at a speed of 28 m/s. If one assumes that the relative velocity at the border of this vortex was ≈28 m/s, this would induce winds of velocity 56 m/s to the south of the vortex and weak winds to the north. Because there was a second vortex of the same kind two days later, these storms present analogies with the secondary vortices observed in the preceding baroclinic jet LES. Such velocities were totally unexpected, and roofs and electric and telephone lines were ill-designed to withstand the forces they exerted. Recall that forces exerted by wind on a body are proportional to the squared velocity when the wake of the body is turbulent. Our constructions could resist winds ≈30 m/s but not 60 m/s, which exert forces four times larger.

Among the differences between the baroclinic-jet simulations and the December 1999 severe storms, Lesieur et al. [175] note pressure troughs measured in the storms, which were not observed in the simulations.

One of the questions posed by Lesieur et al. [175] concerns the requirements needed by a numerical weather-forecast code to capture analogous phenomena correctly and to predict the severe vertical vortex stretching in the thermal fronts. One may wonder whether the hydrostatic approximation is sufficient and if Boussinesq- or anelastic-type approximations of the Navier–Stokes equation should not be preferred. Gravity waves should be filtered out in some way. Meshes smaller than 50 km should be used horizontally, and numerical schemes in the vertical direction should not be too diffusive. The use of LES seems compulsory to obtain secondary instabilities because the latter are dissipated by molecular viscosity in the baroclinic-jet DNS. The question of data assimilation to define the proper initial field is another controversial topic. Developing forecasting tools able to predict the wind velocities associated with such severe storms accurately is of essential importance for populations living in midlatitude regions.

It is difficult to know whether these events are associated with a warming climate. Climatologists stress that polar regions warm faster than equatorial ones, leading to meridional temperature gradients, which should become in the mean weaker rather than stronger, as they certainly were during the December 1999 storms.

8.4 Animations

Animation 8-1: DNS of a baroclic jet at Rossby and Froude numbers of 0.1, showing isosurfaces of vertical relative vorticity (pink, positive; blue, negative). (Film 8-1.mpg; courtesy E. Garnier.)

Bibliography

[1]   Abbà, A., Cercignani, C., Valdaretto, L., and Zanini, P., 1997, LES of turbu-
      lent thermal convection, in *Direct and Large-Eddy Simulation II*, J. P. Chollet,
      L. Kleiser, and P. R. Voke (eds.), Kluwer Academic Publishers, pp. 147–
      56.

[2]   Abbà, A. and Cercignani, C., 1998, private communication.

[3]   Abid, M. and Brachet, M. E., 1993, Numerical characterization of the dynamics
      of vortex filaments in round jets, *Phys. Fluids A* **5**, 2582–4.

[4]   Ackermann, C. and Métais, O., 2001, A modified selective structure function
      subgrid-scale model, *J. Turbulence* **2**, 011.

[5]   Airiau, C., 1994, Stabilité linéaire et faiblement non linéaire d'une couche
      limite laminaire incompressible par un système d'équations parabolisé (PSE)
      PhD thesis, Toulouse.

[6]   André, J. C. and Lesieur, M., 1977, Influence of helicity on high Reynolds
      number isotropic turbulence, *J. Fluid Mech.* **81**, 187–207.

[7]   Antonia, R. A., Teitel, M., Kim, J., and Browne, L. W. B., 1992, Low-Reynolds-
      number effects in a fully developed turbulent channel flow, *J. Fluid Mech.* **236**,
      579–605.

[8]   Antonia, R. A., 1998, private communication.

[9]   Arnal, M. and Friedrich, R., 1993, Large-eddy simulation of a turbulent flow
      with separation, in *Turbulent Shear Flows* VIII, Selected papers from the 8th
      International Symposium on Turbulent Shear Flows, F. Durst, R. Friedrich,
      B. E. Launder, F. W. Schmidt, U. Schumann, J. H. Whitelaw (eds.), Springer-
      Verlag Berlin, New York, pp. 169–87.

[10] Aubry, N., Holmes, P., Lumley, J. L., and Stone, E., 1988, The dynamics of
      coherent structures in the wall region of a turbulent-boundary layer, *J. Fluid
      Mech.* **192**, 115–73.

[11] Balibar, S., 2004, private communication.

[12] Bardina, J., Ferziger, J. H., and Reynolds, W. C., 1980, Improved subgrid model
      for large-eddy simulation, AIAA Paper 80-1357.

[13] Barenblatt, G. and Prostokishin, V., 1993, Scaling laws for fully developed
      turbulent shear flows, *J. Fluid Mech.* **248**, 513–29.

[14] Bartello, P., Métais, O., and Lesieur, M., 1994, Coherent structures in rotating
      three-dimensional turbulence, *J. Fluid Mech.* **273**, 1–29.

[15] Basdevant, C. and Sadourny, R., 1983, Parameterization of virtual scale in numerical simulation of two-dimensional turbulent flows, in *Two-Dimensional Turbulence, J. Mech. Theor. Appl., Suppl.*, R. Moreau (ed.), pp. 243–70.

[16] Basdevant, C. and Philipovitch, T., 1994, On the validity of the 'Weiss criterion' in two-dimensional turbulence, *Physica D* **73**, 17–30.

[17] Batchelor, G. K., 1967, *An Introduction to Fluid Dynamics*, Cambridge University Press.

[18] Batchelor, G. K., 1953, *The Theory of Homogeneous Turbulence*, Cambridge University Press.

[19] Batchelor, G. K., 1951, Pressure fluctuations in isotropic turbulence, *Proc. Cambridge Phil. Soc.* **47**, 359–74.

[20] Beer, A., 2003, Etude par simulation numérique temporelle des effets de compressibilité en couche de mélange plane turbulente, PhD thesis, Grenoble.

[21] Bell, J. and Mehta, R., 1990, Development of a two-stream mixing layer from tripped and untripped boundary layers, *AIAA J.* **28**, 2034–42.

[22] Bernard, S. and Wallace, J., 2002, *Turbulent Flow*, Wiley, 497 pp.

[23] Bernal, L. P. and Roshko, A., 1986, Streamwise vortex structure in plane mixing layer, *J. Fluid Mech.* **170**, 499–525.

[24] Bertolotti, P. and Herbert, T., 1991, Analysis of the linear stability of compressible boundary layers using the PSE, *Theor. Comput. Fluids Dyn.* **3**, 117–24.

[25] Blackadar, A., 1950, The transformation of energy by the large scale eddy stress in the atmosphere, Meteorological Papers, Vol. I, **4**, New York University.

[26] Blumen, W., 1970, Shear-layer instability of an inviscid compressible fluid, *J. Fluid Mech.* **40**, 769–81.

[27] Bogdanoff, D. W., 1983, Compressibility effects in turbulent shear layers, *AIAA J.* **21**, 926–7.

[28] Borue, J. and Orszag, S. A., 1997, Spectra in helical three-dimensional homogeneous isotropic turbulence, *Phys. Rev. E.* **55**, 7005–9.

[29] Brachet, M., 1990, Géométrie des structures à petit échelle dans le vortex de Taylor-Green, *C. R. Acad. Sci. Paris, Ser. B.* **311**, 775–80.

[30] Breuer, M. and Jovicic, J., 2001, Separated flow around a flat plate at high incidence: An LES investigation, *J. Turbulence* **2**, 018.

[31] Briand, E., 1999, Dynamique des structures cohérentes en couche limite transitionnelle et turbulence étudiée par simulation des grandes échelles, PhD thesis, Grenoble.

[32] Browand, F. K. and Troutt, T. R., 1980, A note on spanwise structure in the two-dimensional mixing layer, *J. Fluid Mech.* **93**, 325–36.

[33] Brown, G. and Roshko, A., 1974, On density effects and large structure in turbulent mixing layers, *J. Fluid Mech.* **64**, 775–816.

[34] Broze, G. and Hussain, F., 1996, Transition to chaos in a forced jet: Intermittency, tangent bifurcations and hysteresis, *J. Fluid Mech.* **311**, 37–71.

[35] Cadot, O., Douady, S., and Couder, Y., 1995, Characterization of very low pressure events in 3D turbulence, *Phys. Fluids* **7**, 630.

[36] Cardin, P., 2002, private communication.

[37] Chambres, O., 1997, Analyse expérimentale de la modélisation de la turbulence en couche de mélange supersonique, PhD thesis, Poitiers.

[38] Chasnov, J., 1994, Similarity states of passive-scalar transport in isotropic turbulence, *Phys. Fluids* **6**, 1036–51.

[39] Choi, K. S. and Fujisawa, N., 1993, Possibility of drag reduction using a d-type roughness, *Appl. Sci. Res.* **50**, 315–24.

[40] Choi, H., Moin, P., and Kim. J., 1993, Direct-numerical simulation of turbulent flow over riblets, *J. Fluid Mech.* **225**, 503–39.

[41] Chollet, J. P. and Lesieur, M., 1981, Parameterization of small scales of three-dimensional isotropic turbulence utilizing spectral closures, *J. Atmos. Sci.* **38**, 2747–57.

[42] Chollet, J. P. and Lesieur, M., 1982, Modélisation sous maille des flux de quantité de mouvement et de chaleur en turbulence tridimensionnelle isotrope, *La Météorologie* **29–30**, 183–91.

[43] Chollet, J. P., 1985, Two-point closures as a subgrid scale modelling for large eddy simulations, in *Turbulent Shear Flows IV*, F. Durst and B. Launder (eds.), Lecture Notes in Physics, Springer-Verlag, pp. 62–72.

[44] Chong, M. S., Perry, A. E., and Cantwell, B. J., 1990, A general classification of three-dimensional flow field, *Phys. Fluids A* **2**, 765.

[45] Christensen, T. and Adrian, R., 2000, The velocity and acceleration signatures of small-scale vortices in turbulent channel flow, *J. Turbulence* **3**, 023.

[46] Clark, R. A., Ferziger, J. H., and Reynolds, W. C., 1979, Evaluation of subgrid-scale models using an accurately simulated turbulent flow, *J. Fluid Mech.* **91**, 1–16.

[47] Coantic, M. and Lasserre, J. J., 1999, On pre-dissipative "bumps" and a Reynolds-number dependent spectral parameterization of turbulence, *Eur. J. Mech. B/Fluids* **18**, 1027–47.

[48] Coleman, G. N., Kim, J., and Moser, R. D., 1995, A numerical study of turbulent supersonic isothermal-wall channel flow, *J. Fluid Mech.* **305**, 159–83.

[49] Comte-Bellot, G., 1965, Ecoulement turbulent entre deux parois parallèles, *Publ. Sci. Tech. Ministère de l'Air*, **419**.

[50] Comte-Bellot, G. and Corrsin, S., 1966, The use of a contraction to improve the isotropy of a grid generated turbulence, *J. Fluid Mech.* **25**, 657–82.

[51] Comte, P., Fouillet, Y., Gonze, M. A., Lesieur, M., Métais, O., and Normand, X., 1991, Large-eddy simulations of free-shear layers, in *Turbulence and coherent structures*, O. Métais and M. Lesieur (eds.), Kluwer, pp. 45–73.

[52] Comte, P., Lesieur, M., and Lamballais, E., 1992, Large and small-scale stirring of vorticity and a passive scalar in a 3D temporal mixing layer. *Phys. Fluids A* **4**, 2761–78.

[53] Comte, P., Fouillet, Y., and Lesieur, M., 1992, Simulation numérique des zones de mélange compressibles, *Revue Scientifique et Technique de la Defense*, 3ème trimestre, 43–63.

[54] Comte, P., Silvestrini, J., and Bégou, P., 1998, Streamwise vortices in large eddy simulation of mixing layers, *Eur. J. Mech. B* **17**, 615–37.

[55] Corcos, G. M. and Lin, S. J., 1984, The mixing layer: Deterministic models of a turbulent flow. Part 2. The origin of the three-dimensional motion, *J. Fluid Mech.* **139**, 67–95.

[56] Cousteix, J., 1989, *Turbulence et Couche Limite*, CEPADUES.

[57] Coustols, E. and Cousteix, J., 1994, Performances of riblets in the supersonic regime, *AIAA J.* **32**, 431–33.

[58] Crow, S. C. and Champagne, F. H., 1971, Orderly structure in jet turbulence, *J. Fluid Mech.* **48**, 547–91.

[59] Cucitore, R., Quadrio, M., and Baron, A., 1999, On the effectiveness and lim-
      itations of local criteria for the identification of a vortex, *Eur. J. Mech. B* **18**,
      261–82.

[60] Danet, A., 2001, Influence des conditions amont sur l'écoulement derrière une
      marche par la simulation des grandes échelles, PhD thesis, Grenoble.

[61] David, E., 1993, *Modélisation des ecoulements compressibles et hypersoniques:
      une approche instationnaire*, PhD thesis, Grenoble.

[62] Davidson, P. A., 2000, Was Loitsyansky correct? A review of the arguments,
      J. Turbulence **1**, 006.

[63] Deardorff, J. W., 1970, A numerical study of three-dimensional turbulent chan-
      nel flow at large Reynolds number, *J. Fluid Mech.* **41**, 453–80.

[64] Deardorff, J. W., 1972, Numerical investigation of neutral and unstable planetary
      boundary layers, *J. Atmos. Sci.* **29**, 91–115.

[65] Deardorff, J. W., 1973, The use of subgrid transport equations in a three-
      dimensional model of atmospheric turbulence, *J. Fluids Eng.* **95**, 429–38.

[66] Deardorff, J. W., 1974, Three-dimensional numerical study of turbulence in an
      entraining mixed layer, *Boundary Layer Meteorol.* **7**, 199–226.

[67] De Bisschop, J., Chambres, O., and Bonnet, J. P., 1994, Velocity field charac-
      teristics in supersonic mixing layers, *Exp. Therm. Fluid Sci.* **9**, 147–55.

[68] Dedebant, T. and Wehrle, P., 1938, Sur les équations aux valeurs probables d'un
      fluide turbulent, *C.R. Acad. Sci.* **206**, 1790–91.

[69] Delcayre, F., 1999, Etude par simulation des grandes échelles d'un écoulement
      décollé: la marche descendante, PhD thesis, Grenoble.

[70] Domaradzki, J. A., Metcalfe, R. W., Rogallo, R. S., and Riley, J. J., 1987, Analysis
      of subgrid-scale eddy viscosity with the use of results from direct numerical
      simulations, *Phys. Rev. Lett.* **58**, 547–50.

[71] Domaradzki, J. A. and Horiuti, K., 2001, Similarity modeling on an expanded
      mesh applied to rotating turbulence, *Phys. Fluids.* **13**, 3510–12.

[72] Domaradzki, J. A. and Adams, N. A., 2002, Direct modelling of subgrid scales
      of turbulence in large eddy simulations, *J. Turbulence* **3**, 024.

[73] Doris, L., 2001, Simulation des grandes échelles du développement spatial
      d'une couche de mélange turbulente compressible, PhD thesis, Paris.

[74] Drazin, P. G. and Reid, W. H., 1981, *Hydrodynamic Stability*, Cambridge Uni-
      versity Press.

[75] Drobniak, S. and Klajny, R., 2002, Coherent structures of free acoustically
      stimulated jet, *J. Turbulence* **3**, 001.

[76] Dubief, Y. and Comte, P., 1997, Large-eddy simulation of a boundary layer
      flow passing over a groove, in *Turbulent Shear Flows 11*, Grenoble, France,
      pp. 1-1–1-6.

[77] Dubief, Y., 2000, Simulation des grandes échelles de la turbulence de la région
      de proche paroi et des écoulements décollés, PhD thesis, Grenoble.

[78] Dubief, Y. and Delcayre, F., 2000, On coherent-vortex identification in turbu-
      lence, *J. Turbulence* **1**, 011.

[79] Ducros, F., 1995, Simulations numériques directes et des grandes échelles de
      couches limites compressibles, PhD thesis, Grenoble.

[80] Ducros, F., Comte, P., and Lesieur, M., 1995, Direct and large-eddy simulations
      of a supersonic boundary layer, in *Selected Proceedings of Turbulent Shear
      Flows 9*, pp. 283–300, Springer.

[81] Ducros, F., Comte, P., and Lesieur, M., 1996, Large-eddy simulation of transition to turbulence in a boundary layer developing spatially over a flat plate, *J. Fluid Mech.* **326**, 1–36.

[82] Ducros, F., Nicoud, F., and Schonfeld, T., 1997, Large-eddy simulation of compressible flows on hybrid meshes, in *Proceedings of the 11th Symposium on Turbulent Shear Flows*, **3**, 28-1–28-6.

[83] Durbin, P. and Pettersson Reif Chichester, B., 2001, *Statistical Theory and Modeling for Turbulent Flows*, Wiley, 302 pp.

[84] El Hady, N. M. and Zang T. A., 1995, Large-eddy simulation of nonlinear evolution and breakdown to turbulence in high-speed boundary layers, *Theor. Comput. Fluid Dyn.* **7**, 217–40.

[85] Elliott, G. S. and Samimy, M., 1990, Compressibility effects in free shear layers, *Phys. Fluids A* **2**, 1230–31.

[86] Erlebacher, G., Hussaini, M. Y., Speziale, C. G., and Zang, T. A., 1987, ICASE Rep., 87–20.

[87] Erlebacher, G., Hussaini, M. Y., Speziale, C. G., and Zang, T. A., 1992, Towards a large-eddy simulation of compressible turbulent flows, *J. Fluid Mech.* **238**, 155–85.

[88] Eyink, G. and Thomson, D., 2000, Free decay of turbulence and breakdown of self-similarity, *Phys. Fluids* **12**, 477–79.

[89] Falkovich, G., 1994, Bottleneck phenomenon in developed turbulence, *Phys. Fluids* **6**, 1411–14.

[90] Fallon, B., Lesieur, M., Delcayre, F., and Grand, D., 1997, Large-eddy simulations of stable-stratification effects upon a backstep flow, *Eur. J. Mech. B/Fluids* **16**, 525–644.

[91] Fauve, S., Laroche, C., and Castaing, B., 1993, Pressure fluctuations in swirling turbulent flows, *J. Phys. II* **3**, 271–8.

[92] Favre, A., 1958, Equations statistiques des gaz turbulents, *C. R. Acad. Sci. Paris*, **246**: masse, quantité de mouvement, 2576–9; énergie cinétique, énergie cinétique du mouvement macroscopique, énergie cinétique de la turbulence, 2839–42; enthalpies, entropie, températures, 3216–19.

[93] Favre, A., 1965, Equations des gaz turbulents compressibles, *J. Mech.* **4**, 361.

[94] Favre, A., Kovasznay, L., Dumas, R., Gaviglio, J., and Coantic, 1976, *La turbulence en mécanique des fluides: bases théoriques et expérimentales, méthodes statistiques*, Gauthier-Villars.

[95] Favre, A., 1983, Turbulence: space-time statistical properties and behavior in supersonic flows, *Phys. Fluids* **26**, 2851–63.

[96] Flores, C., 1993, Etude numérique de l'influence d'une rotation sur les écoulements cisaillés libres, PhD thesis, Grenoble.

[97] Forster, D., Nelson, D. R., and Stephen, M. J., 1977, Large distance and long time properties of a randomly stirred field, *Phys. Rev. A* **16**, 732–49.

[98] Fouillet, Y., 1992, Contribution à l'étude par expérimentation numérique des écoulements cisaillés libres. Effets de compressibilité, PhD thesis, Grenoble.

[99] Fournier, J. D., 1977, Quelques méthodes systématiques de développement en turbulence homogène, PhD thesis, Nice.

[100] Fournier, J. D. and Frisch, U., 1983, Remarks on the renormalization group in statistical fluid dynamics, *Phys. Rev. A* **28**, 1000–1002.

[101] Freund, J. B., 1999, Acoustic sources in a turbulent jet: A direct numerical simulation study, AIAA Paper 99-1858.

[102] Freund, J. B., Lele, S. K., and Moin, P., 2000, Compressibility effects in a turbulent annular mixing layer. Part 1. Turbulence and growth rate, *J. Fluid Mech.* **421**, 229–67.

[103] Frisch, U., 1995, *Turbulence, the Legacy of A.N. Kolmogorov*, Cambridge University Press.

[104] Fureby, C. and Grinstein, F., 1999, Monotonically integrated large eddy simulation of free shear flows, *AIAA J.* **37**, 544–56.

[105] Fureby, C. and Grinstein, F., 2002, Large eddy simulation of high Reynolds number free and wall bounded flows, *J. Comput. Phys.* **181**, 1–30.

[106] Garnier, E., Métais, O., and Lesieur, M., 1998, Synoptic and frontal-cyclone scale instabilities in baroclinic jet flows, *J. Atmos. Sci.* **55**(8), 1316–35.

[107] Gavrilakis, S., 1992, Numerical simulation of low Reynolds number turbulent flow through a straight square duct, *J. Fluid Mech.* **244**, 101–29.

[108] Germano, M., Piomelli, U., Moin, P., and Cabot, W., 1991, A dynamic subgrid-scale eddy-viscosity model, *Phys. Fluids A* **3**(7), 1760–65.

[109] Germano, M., 1992, Turbulence, the filtering approach, *J. Fluid Mech.* **238**, 325–36.

[110] Geurts, B., 1997, Inverse modelling for large-eddy simulation, *Phys. Fluids* **9**, 3585–7.

[111] Geurts, B., 2003, *Elements of Direct and Large-Eddy Simulation*, Edwards.

[112] Goebel, S. G. and Dutton, J. C., 1991, Experimental study of compressible turbulent mixing layers, *AIAA J.* **29**, 538–46.

[113] Gonze, M. A., 1993, Simulation numérique des sillages en transition à la turbulence, PhD thesis, Grenoble.

[114] Gottlieb, D. and Turkel, E., 1976, Dissipative two-four methods for time-dependant problems, *Math. Comp.* **30**, 703–23.

[115] Grinstein, F., Young, T., Gutmark, E., Li, G., Hsiao, G., and Mongia, H., 2002, Flow dynamics in a swirl combustor, *J. Turbulence* **3**, 030.

[116] Gurbatov, S., Simdyankin, S., Aurell, E., Frisch, U., and Toth, G., 1997, On the decay of Burgers turbulence, *J. Fluid Mech.* **344**, 339–74.

[117] Hanjalic, K. and Kenjeres, S., 2000, Reorganization of turbulence structure in magnetic Rayleigh-Bénard convection: A T-RANS study, *J. Turbulence* **1**, 008.

[118] Hauët, G., 2003, Contrôle de la turbulence par simulation des grandes échelles en transport supersonique, PhD thesis, Grenoble.

[119] Hébrard, J., Salinas-Vasquez, M., and Métais, O., 2004, Large-eddy simulation of turbulent duct flow: heating and curvature effects, *Int. J. Heat and Fluid Flow* **25**, 569–80.

[120] Heisenberg, W., Zur statistischen Theorie der Turbulenz, *Z. Phys.* **124**, 628–57.

[121] Herbert, T., 1988, Secondary instability of boundary layers, *Annu. Rev. Fluid Mech.* **20**, 487–526.

[122] Herbert, T., 1997, Parabolized stability equations *Annu. Rev. Fluid Mech.* **29**, 245–83.

[123] Hesselberg, 1925, Die Gesetze der ausgeglichenen atmosphärischen Bewegangen, *Beitr. Phys. Atmos.* **12**, 141–60.

[124] Hill, R., 1996, Pressure-velocity statistics in isotropic turbulence, *Phys. Fluids* **8**, 3085–93.

[125] Hoffmann, P. H, Muck, K. C, and Bradshaw, P., 1985, The effect of concave surface curvatures on turbulent boundary layers, *J. Fluid Mech.* **161**, 371–403.

[126] Holmes, P., Lumley, J., and Berkooz, G., 1998, *Turbulence, Coherent Structures, Dynamical Systems and Symmetry*, Series: Cambridge Monographs on Mechanics, Cambridge University Press, p. 438.

[127] Horiuti, K., 2001, Rotational transformation and geometrical correlation of SGS models, in *Modern Simulation Strategies for Turbulent Flow*, B. Geurts (ed.), Edwards, pp. 123–40.

[128] Hunt, I. A., and Joubert, P. N., 1979, Effects of small streamline curvature on turbulent duct flow, *J. Fluid Mech.* **91**, 633.

[129] Hunt, J., Wray, A., and Moin, P., 1988, Eddies, stream, and convergence zones in turbulent flows. Center for Turbulence Research Rep., **CTR-S88**, 193.

[130] Hunt, J., 1998, Prévision déterministe et statistique de l'environnement et de la turbulence, in *Turbulence et déterminisme*, Collection Grenoble-Sciences, M. Lesieur (ed.), EDP-Springer, pp. 17–48.

[131] Hussein, H. J., Capp, S. P., and George, W. K., 1994, Velocity measurements in high-Reynolds number, momentum-conserving, axisymmetric, turbulent jet, *J. Fluid Mech.* **258**, 31–75.

[132] Issa, R., 2001, Refroidissement des chambres de combustion de moteurs-fusées, Master's thesis, Grenoble.

[133] Jeong, J. and Hussain, F., 1995, On the identification of a vortex, *J. Fluid Mech.* **285**, 69–94. See also Jeong, J., Hussain, J., Schoppa, W., and Kim, J., 1997, Coherent structures near the wall in a turbulent channel flow, *J. Fluid Mech.* **332**, 185–214.

[134] Jimenez, J., 1990, Transition to turbulence in two-dimensional Poiseuille flow, *J. Fluid Mech.* **218**, 265–97.

[135] Jimenez, J. and Wray, A. A., 1998, On the characteristics of vortex filaments in isotropic turbulence, *J. Fluid Mech.* **373**, 255–85.

[136] Jimenez, J., 2000, Turbulence, in *Perspectives in Fluid Mechanics*, G. K. Batchelor, H. K. Moffatt, and M. G. Woster (eds.), Cambridge University Press, pp. 231–83.

[137] Johnston, J. P., Halleen, R. M., and Lezius, D. K., 1972, Effects of spanwise rotation on the structure of two-dimensional fully developed turbulent channel flow, *J. Fluid Mech.* **56**, 553–57.

[138] Jovic, S. and Driver, M., 1994, Backward-facing step measurement at low Reynolds number $Re_h = 5000$, Ames Research Center, NASA Technical Memorandum, 108807.

[139] Kachanov, Y. and Levchenko, V., 1984, The resonant interaction of disturbances at laminar-turbulent transition in a boundary layer, *J. Fluid Mech.* **138**, 209–47.

[140] Kenjeres, S. and Hanjalic, K., Combined effects of terrain orography and thermal stratification on pollutant dispersion in a town valley: A T-RANS simulation, *J. Turbulence* **3**, 026.

[141] Kim, J., 1983, The effect of rotation on turbulence structure, in *Proceedings of the 4th Symposium on Turbulent Shear Flows, Karlsruhe*, 6.14–6.19.

[142] Kim, J., Moin, P., and Moser, R., 1987, Turbulent statistics in fully developed channel flow at low Reynolds number, *J. Fluid Mech.* **177**, 133–66.

[143] Klebanoff, P. S., Tidstrom, K. D., and Sargent, L. M., 1962, The three-dimensional naure of boundary layer instability, *J. Fluid Mech.* **12**, 1–34.

[144] Kline, S. J., Reynolds, W. C., Schraub, F. A., and Runstadler, P. W., 1967, The structure of turbulent-boundary layers, *J. Fluid Mech.* **30**, 741–73.

[145] Kolmogorov, A. N., 1941, The local structure of turbulence in incompressible viscous fluid for very large Reynolds numbers, *Dokl. Akad. Nauk SSSR* **30**, 301–305.

[146] Kolmogorov, A. N., 1941, On degeneration of isotropic turbulence in an incompressible viscous liquid, *Dokl. Akad. Nauk. SSSR* **31**, 538–41.

[147] Kraichnan, R. H., 1976, Eddy viscosity in two and three dimensions, *J. Atmos. Sci.* **33**, 1521–36.

[148] Kristoffersen, R. and Andersson, H. I., 1993, Direct simulations of low-Reynolds-number turbulent flow in a rotating channel, *J. Fluid Mech.* **256**, 163–97.

[149] Kuroda, A., 1990, Direct-numerical simulation of Couette-Poiseuille flows, PhD thesis, Tokyo.

[150] Kusek, S. M., Corke, T. C., and Reisenthel, R., 1990, Seeding of helical modes in the initial region of an axisymmetric jet, *Exp. Fluids* **10**, 116–24.

[151] Lamballais, E., 1995, Simulations numériques de la turbulence dans un canal plan tournant, PhD thesis, Grenoble.

[152] Lamballais, E., Lesieur, M., and Métais, O., 1996, Effects of spanwise rotation on the vorticity stretching in transitional and turbulent channel flow, *Int. J. Heat Fluid Flow* **17**, 324–32.

[153] Lamballais, E., Lesieur, M., and Métais, O., 1997, Probability distribution functions and coherent structures in a turbulent channel, *Phys. Rev. E.* **56**, 6761–6.

[154] Lamballais, E., Métais, O., and Lesieur, M., 1998, Spectral-dynamic model for large-eddy simulations of turbulent rotating channel flow, *Theor. Comput. Fluid Dyn.* **12**, 149–77.

[155] Landau, L. and Lifchitz, E., 1971, *Fluid Mechanics*, Mir.

[156] Larchevêque, M., 1990, Pressure fluctuations and Lagrangian accelerations in two-dimensional incompressible isotropic turbulence, *Eur. J. Mech. B* **9**, 109–28.

[157] Lasheras, J. C., Lecuona, A., and Rodriguez, P., 1991, Three dimensionnal structure of the vorticity field in the near region of laminar co-flowing forced jets, in *The Global Geometry of Turbulence*, J. Jimenez (ed.), Plenum.

[158] Laufer, J., 1950, Investigation of turbulent flow in a two-dimensional channel, NACA Rep. 1053.

[159] Le, H., Moin, P., and Kim, J., 1997, Direct numerical simulation of turbulent flow over a backward-facing step, *J. Fluid Mech.* **330**, 349–74.

[160] Lechner, R., Sesterhenn, J., and Friedrich, R., 2001, Turbulent supersonic channel flow, *J. Turbulence* **2**, 001.

[161] Lee, M. and Reynolds, W. C., 1985, Bifurcating and blooming jets at high Reynolds number, in *Fifth Symposium on Turbulent Shear Flows*, Ithaca, New York, 1.7–1.12.

[162] Leith, C. E., 1990, Stochastic backscatter in a subgrid-scale model: Plane shear mixing layer, *Phys. Fluids A* **2**, 297–300.

[163] Leonard, A., 1974, Energy cascade in large eddy simulations of turbulent fluid flows, *Adv. Geophys. A* **18**, 237–48.

[164] Lesieur, M. and Schertzer, D., 1978, Amortissement auto similaire d'une turbulence à grand nombre de Reynolds, *J. Mech.* **17**, 609–46.

[165] Lesieur, M. and Rogallo, R., 1989, Large-eddy simulation of passive-scalar diffusion in isotropic turbulence, *Phys. Fluids A* **1**, 718–22.

[166] Lesieur, M., Yanase, S., and Métais, O., 1991, Stabilizing and destabilizing effects of a solid-body rotation on quasi-two-dimensional shear layers, *Phys. Fluids A* **3**, 403–7.

[167] Lesieur, M., 1994, *La turbulence*, Collection Grenoble-Sciences, EDP-Springer.

[168] Lesieur, M., and Métais, O., 1996, New trends in large-eddy simulations of turbulence, *Annu. Rev. Fluid Mech.* **28**, 45–82.

[169] Lesieur, M., 1997, Recent approaches in large-eddy simulations of turbulence, in *New Tools in Turbulence Modelling*, O. Métais and J. Ferziger (eds.), Les Éditions de Physique, Springer-Verlag, pp. 1–28.

[170] Lesieur, M., 1997, *Turbulence in Fluids,* 3rd Revised and Enlarged Ed., Kluwer, 515 pp.

[171] Lesieur, M., Ossia, S., and Métais, O., 1999, Infrared pressure spectra in two- and three-dimensional isotropic incompressible turbulence, *Phys. Fluids* **11**, 1535–43.

[172] Lesieur, M., Comte, P., Dubief, Y., Lamballais, E., Métais, O., and Ossia, S., 1999, From two-point closures of isotropic turbulence to LES of shear flows, *Flow Turbulence Comb.* **63**, 247–67.

[173] Lesieur, M., Delcayre, F., and Lamballais, E., 2000, Spectral eddy-viscosity based LES of shear and rotating flows, in *Developments in Geophysical Turbulence*, R. Kerr and Y. Kimura (eds.), Kluwer, pp. 235–52.

[174] Lesieur, M. and Ossia, S., 2000, 3D isotropic turbulence at very high Reynolds numbers: EDQNM study, *J. Turbulence* **1**, 007.

[175] Lesieur, M., Métais, O., and Garnier, E., 2000, Baroclinic instability and severe storms, *J. Turbulence* **1**, 02.

[176] Lesieur, M., Bégou, P., Briand, E., Danet, A., Delcayre, F., and Aider, J. L., 2003, Coherent-vortex dynamics in large-eddy simulations of turbulence, *J. Turbulence* **4**, 016.

[177] Leslie, D. C. and Quarini, G. L., 1979, The application of turbulence theory to the formulation of subgrid modelling procedures, *J. Fluid Mech.* **91**, 65–91.

[178] Lessen, M., Fox, J. A., and Zien, H. M., 1965, On the inviscid stability of the laminar mixing of two parallel streams of a compressible fluid, *J. Fluid Mech.* **23**, 355–67.

[179] Lessen, M., Fox, J. A., and Zien, H. M., 1966, Stability of the laminar mixing of two parallel streams with respect to supersonic disturbances, *J. Fluid Mech.* **25**, 737–42.

[180] Lezius, D. K. and Johnston, J. P., 1976, Roll-cell instabilities in rotating laminar and turbulent channel flows, *J. Fluid Mech.* **77**, 153–75.

[181] Liepmann, D. and Gharib, M., 1992, The role of streamwise vorticity in the near-field entrainment of round jets, *J. Fluid Mech.* **245**, 643–68.

[182] Lilly, D. K., 1962, On the numerical simulation of buoyant convection, *Tellus* **14**, 148–72.

[183] Lilly, D. K., 1966, On the application of the eddy-viscosity concept in the inertial subrange of turbulence, NCAR manuscript 123.

[184] Lilly, D. K., 1967, The representation of small-scale turbulence in numerical simulation experiments, in *Proc, IBM Sci. Comput. Symp. on Environ. Sci.*, IBM Form 320-1951, pp. 195–210.

[185] Lilly, D. K., 1987, Lecture notes by Douglas Lilly, in *Lecture Notes on Turbulence*, J. R. Herring and J. C. McWilliams (eds.), World Scientific, pp. 171–218.

[186] Lilly, D. K., 1992, A proposed modification of the Germano subgrid-scale closure method, *Phys. Fluids A* **4** (3), 633–35.

[187] Lilly D .K., 1996, Numerical simulation and prediction of atmospheric convection, in *Computational Fluids Dynamics, Les Houches; Session LIX*, 1993, M. Lesieur, P. Comte, J. Zin-Justin (eds.), Elsevier, pp. 325–74.

[188] Liu, S., Meneveau, C., and Katz, J., 1994, On the properties of similarity subgrid-scale models as deduced from measurements in turbulent jet, *J. Fluid Mech.* **275**, 83–119.

[189] Longmire, E. K. and Duong, L. H., 1996, Bifurcating jets generated with stepped and sawtooth nozzles, *Phys. Fluids*, **8**(4), 978–92.

[190] Lumley, J., 1970, *Stochastic Tools in Turbulence*, Academic Press.

[191] Lumley, J. and Newman, G., 1977, The return to isotropy of homogeneous turbulence, *J. Fluid Mech.* **82**, 161–78.

[192] Lund, T., Wu, X., and Squires, K. D., 1998, Generation of turbulent inflow data for spatially-developing boundary-layer simulations, *J. Comput. Phys.* **140**, 233–58.

[193] MacCormack, R., 1969, The effect of viscosity in hypervelocity impact cratering, AIAA Paper 69-354.

[194] Mandelbrot, B., 1977, *The Fractal Geometry of Nature*, Freeman.

[195] Mankbadi, R., 2001, Computational aeroacoustics, in *New Trends in Turbulence*, M. Lesieur, A. Yaglom, and F. David (eds.), EDP-Springer, 250.

[196] Martin, J. E. and Meiburg, E., 1991, Numerical investigation of three-dimensionally evolving jets subject to axisymmetric and azimuthal perturbations, *J. Fluid Mech.* **230**, 271–318.

[197] Martin, J. E. and Meiburg, E., 1992, Numerical investigation of three-dimensionally evolving jets under helical perturbations, *J. Fluid Mech.* **243**, 457–87.

[198] Mas, D., 2000, Rayonnement acoustique d'une cavité rectangulaire soumise à un écoulement turbulent, PhD thesis, Grenoble.

[199] Mason, P., 1988, Large-eddy simulation of the convective atmospheric boundary layer, *J. Atmos. Sci.* **46**, 1492–516.

[200] Mc Comb, W. D., 1990, The physics of fluid turbulence, Oxford University Press.

[201] Meneveau, C., Lund, T. S., and Cabot, W. H., 1996, A Lagrangian dynamic subgrid-scale model of turbulence, *J. Fluid Mech.* **319**, 353–86.

[202] Meneveau, C. and Katz, J., 2000, Scale-invariance and turbulence models for large-eddy simulations, *Annu. Rev. Fluid Mech.* **32**, 1–32.

[203] Mestayer, P., Chollet, J. P., and Lesieur, M., 1984, Inertial subrange of velocity and scalar variance spectra in high Reynolds number three-dimensional turbulence, in *Turbulence and Chaotic Phenomena in Fluids*, T. Tatsumi (ed.), North-Holland, pp. 285–88.

[204] Métais, O. and Lesieur, M., 1986, Statistical predictability of decaying turbulence, *J. Atmos. Sci.* **43**, 857–70.

[205] Métais, O. and Lesieur, M., 1992, Spectral large-eddy simulations of isotropic and stably-stratified turbulence. *J. Fluid Mech.* **239**, 157–94.

[206] Métais, O., Flores, C., Yanase, S., Riley, J. J., and Lesieur, M., 1995, Rotating free shear flows. Part 2: Numerical simulations, *J. Fluid Mech.* **293**, 41–80.

[207] Métais, O., 2003, Large-eddy simulations of the heat exchanges in turbulent ducts, in *Turbulence, Heat and Mass Transfer 4*, K. Hanjalic, Y. Nagano, and M. Tummers (eds.), Begell House, pp. 25–35.

[208] Michalke, A., 1964, On the inviscid instability of the hyperbolic tangent velocity profile, *J. Fluid Mech.* **19**, 543–56.

[209] Michalke, A. and Hermann, G., 1982, On the inviscid instability of a circular jet with external flow, *J. Fluid Mech.* **114**, 343–59.

[210] Moin, P. and Kim, J., 1982, Numerical investigation of turbulent channel flow, *J. Fluid Mech.* **118**, 341–77.

[211] Moin P., Squires K., Cabot W., and Lee S., 1991, A dynamic subgrid-scale model for compressible turbulence and scalar transport, *Phys. Fluids A* **3**(11), 2746–57.

[212] Monin, A. S. and Yaglom, A. M., 1975, *Statistical Fluid Mechanics: Mechanics of Turbulence*, Vol. 2, M.I.T. Press.

[213] Monkewitz, P. A. and Pfizenmaier, E., 1991, Mixing by "side jets" in strongly forced and self-excited round jets, *Phys. Fluids A* **3**(5), 1356–61.

[214] Münch, C., Hébrard, J., and Métais, O., 2003, Large-eddy simulations of curved and S-shape squared ducts, in *Direct and Lare-Eddy Simulation 5*, Munich, 27–29 August 2003.

[215] Muchinski, A., 1996, A similarity theory of locally homogeneous and isotropic turbulence generated by a Smagorinsky-type LES, *J. Fluid Mech.* **325**, 239–60.

[216] Nagano, Y. and Hattori, H., 2002, An improved turbulence model for rotating shear flows, *J. Turbulence* **3**, 006.

[217] Nakabayashi, K. and Kitoh, O., 1996, Low Reynolds number fully developed two-dimensional turbulent channel flow with system rotation, *J. Fluid Mech.* **315**, 1–29.

[218] Neu, J. C., 1984, The dynamics of stretched vortices, *J. Fluid Mech.* **143**, 253–76.

[219] Nicoud, F. and Ducros, F., 1999, Subgrid-scale modelling based on the square of the velocity gradient tensor, *Flow Turbulence Comb.* **62**, 183–200.

[220] Normand, X., 1990, Transition à la turbulence dans les écoulements cisaillés libres et pariétaux, PhD thesis, Grenoble.

[221] Normand, X. and Lesieur, M., 1992, Direct and large-eddy simulation of transition in the compressible boundary layer, *Theor. Comput. Fluid Dyn.* **3**, 231–52.

[222] Ohkitani, K., 1993, Eigenvalue problems in three-dimensional Euler flows, *Phys. Fluids A* **5**, 2570–72.

[223] Orlanski, I., 1976, A simple boundary condition for unbounded hyperbolic flows, *J. Comput. Phys.* **21**, 251–69.

[224] Orszag, S. A., 1970, Analytical theories of turbulence, *J. Fluid Mech.* **41**, 363–86.

[225] Orszag, S. A., 1977, Statistical theory of turbulence, in *Fluid Dynamics 1973*, Les Houches Summer School of Theoretical Physics, R. Balian and J. L. Peube (eds.), Gordon and Breach, pp. 237–374.

[226] Ossia, S. and Lesieur, M., 2000, Energy backscatter in large-eddy simulations of three-dimensional incompressible isotropic turbulence, *J. Turbulence* **1**, 010.

[227] Ossia, S., 2000, La dynamique des échelles infrarouges en turbulence isotrope incompressible, PhD thesis, Grenoble.

[228] Padilla-Barbosa, J. and Métais, O., 2000, Large-eddy simulations of deep-ocean convection: analysis of the vorticity dynamics, *J. Turbulence* **1**, 009.

[229] Pantano, C. and Sarkar, S., 2002, A study of compressibility effects in the high-speed turbulent shear layer using direct simulation, *J. Fluid Mech.* **451**, 329–71.

[230] Papamoschou, D. and Roshko, A., 1988, The compressible turbulent shear layer: An experimental study, *J. Fluid Mech.* **197**, 453–77.

[231] Parekh, D.E., Leonard, A., and Reynolds, W. C., 1988, Bifurcating jets at high Reynolds number. Air Force Office of Scientific Research Contractor Rep. AF-F49620-84-K-0005 and AF-F49620-86-K-0020.

[232] Pearson, B. R., Elavarasan, R., and Antonia, R. A., 1997, The response of a turbulent boundary layer to a square groove, *J. Fluids Eng.* **199**, 466–69.

[233] Pedley, T. J., 1969, On the stability of viscous flow in a rapidly rotating pipe, *J. Fluid Mech.* **35**, 97–115.

[234] Petersen, R. A., 1978, Influence of wave dispersion on vortex pairing in a jet, *J. Fluid Mech.* **89**, 469–95.

[235] Pierrehumbert, R. T. and Widnall, S. E., 1982, The two- and three-dimensional instabilities of a spatially periodic shear layer, *J. Fluid Mech.* **114**, 59–82.

[236] Piomelli, U., 1993, High Reynolds number calculations using the dynamic subgrid-scale stress model, *Phys. Fluids A* **5**(6), 1484–90.

[237] Piomelli, U. and Liu, J., 1995, Large-eddy simulation of rotating channel flows using a localized dynamic model, *Phys. Fluids A* **7**(4), 839–48.

[238] Piomelli, U., Balaras, E., and Pascarelli, A., 2000, Turbulent structures in accelerating boundary layers, *J. Turbulence* **1**, 001.

[239] Piomelli, U. and Balaras, B., 2002, Wall-layer models for large-eddy simulations, *Annu. Rev. Fluid Mech.* **34**, 349–74.

[240] Poinsot, T. J. and Lele, S. K., 1992, Boundary conditions for direct simulations of compressible viscous flows, *J. Comput. Phys.* **101**, 104–29.

[241] Reynolds, W. C., Parekh, D. E., Juvet, P. J. D., and Lee, H. J. D., 2003, Bifurcating and blooming jets, *Annu. Rev.* **35**, 295–315.

[242] Reynolds, W. and Hussain, A., 1971, The mechanism of an organized wave in turbulent shear flow. Part 3: Theoretical models and comparison with experiments, *J. Fluid Mech.* **54**, 263–88.

[243] Richardson, L. F., 1922, *Weather Prediction by Numerical Process*, Cambridge University Press.

[244] Rodi, W., Ferziger, J. H., Breuer, M., and Pourquié, M., 1997, Status of large eddy simulation: Results of a workshop, Workshop on LES of Flows Past Bluff Bodies (Rottach-Egern, Tegernsee, 1995), *J. Fluids Eng.* **119**, 248–62.

[245] Rogers, M. and Moser, R., 1994, Direct simulation of a self-similar turbulent mixing layer, *Phys. Fluids A* **6**, 903–24.

[246] Rütten, F., Meinke, M., and Schröder, W., 2001, Large-eddy simulations of 90° pipe bend flows, *J. Turbulence* **2**, 003.

[247] Saffman P. G., 1967, The large-scale structure of homogeneous turbulence, *J. Fluid Mech.* **27**, 581–93.

[248] Sagaut, P., 2001, *Large-Eddy Simulations for Incompressible Flows: An Introduction*, Springer, 319 pp.

[249] Salinas-Vazquez, M., 1999. Simulations des grandes échelles des écoulements turbulents dans les canaux de refroidissement des moteurs fusée, PhD thesis, Grenoble.

[250] Salinas-Vazquez, M. and Métais, O., 1999, Large-eddy simulation of a square duct with a heat flux, in *Direct and Large-Eddy Simulation III*, P. R. Voke, N. D. Sandham, and L. Kleiser (eds), Kluwer Academic Publishers, pp. 13–24.

[251] Salinas-Vazquez, M. and Métais, O., 2002, Large eddy simulation of the turbulent flow through a heated square duct, *J. Fluid Mech.* **453**, 201–38.

[252] Samimy, M. and Elliott, G. S., 1990, Effects of compressibility on the characteristics of free shear layers, *AIAA J.* **28**, 439–45.

[253] Sandham, N. D. and Reynolds, W. C., 1989, The compressible mixing layer: Linear theory and direct simulation, AIAA Paper 89-0371.

[254] Sandham, N. D. and Reynolds, W. C., 1991, Three-dimensional simulations of large eddies in the compressible mixing layer, *J. Fluid Mech.* **224**, 133–58.

[255] Sankaran, V. and Menon, S., 2002, LES of spray combustion in swirling flows, *J., Turbulence* **3**, 011.

[256] Saric, W. S., 1994, Görtler vortices, *Annu. Rev. Fluid Mech.* **26**, 379–409.

[257] Schlichting, H., 1979, *Boundary-Layer Theory*, McGraw-Hill.

[258] Schmidt, H. and Schumann, U., 1989, Coherent structure in the convective boundary layer derived from large-eddy simulations, *J. Fluid Mech.* **200**, 511–62.

[259] Schumann, U., 1975, Subgrid scale model for finite-difference simulations of turbulent flows in plane channels and annuli, *J. Comput. Phys.* **18**, 376–404.

[260] Schumann, U., 1994, On relations between constants in homogeneous turbulence models and Heisenberg's spectral model, *Beitr. Phys. Atmos.* **67**, 141–47.

[261] Settles, G. S., Fitzpatrick, T. J., and Bogdonoff, S. M., 1979, Detailed study of attached and separated compression corner flowfields in high-Reynolds-number supersonic flow, *AIAA J.* **17**(6), 579–85.

[262] She, Z. S., Jackson, E., and Orszag, S. A., 1990, *Nature* **344**, 226.

[263] Siggia, E. D., 1981, Numerical study of small-scale intermittency in three-dimensional turbulence, *J. Fluid Mech.* **107**, 375–406.

[264] Silva, C., 2001, The role of coherent structures in the control and interscale interactions of round, plane and coaxial jets, PhD thesis, Grenoble.

[265] Silva, C. and Métais, O., 2002, Vortex control of bifurcating jets: A numerical study, *Phys. Fluids* **14**, 3798–819.

[266] Silveira-Neto, A., Grand, D., Métais, O., and Lesieur, M., 1993, A numerical investigation of the coherent structures of turbulence behind a backward-facing step, *J. Fluid Mech.* **256**, 1–55.

[267] Silvestrini, J., Simulation des grandes échelles des zones de mélange: application à la propulsion solide des lanceurs spatiaux, PhD thesis, Grenoble.

[268] Silvestrini, J., Lamballais, E., and Lesieur, M., 1998, Spectral-dynamic model for LES of free and wall shear flows, *Int. J. Heat Fluid Flow*, **19**, 492–504.

[269] Smagorinsky, J., 1963, General circulation experiments with the primitive equations, *Mon. Weath. Rev.* **91**(3), 99–164.

[270] Smits, L. and Dussauge, J. P., 1996, *Turbulent Shear Layers in Supersonic Flows*, American Institute of Physics.

[271] Somméria, G., 1976, Three-dimensional simulation of turbulent processes in an undisturbed trade wind boundary layer, *J. Atmos. Sci.* **33**, 216–41.

[272] Somméria, J., 2001, Two-dimensional turbulence, in *New Trends in Turbulence*, M. Lesieur, A. Yaglom, and F. David (eds.), EDP-Springer, 554 pp.

[273] Spalart, P. R., 1988, Direct simulation of a turbulent boundary layer up to $R_\theta = 1410$, *J. Fluid Mech.* **187**, 61–98.

[274] Stalp, S., Skrbeck, L., and Donnelly, J., 1999, Decay of grid turbulence in a finite channel, *Phys. Rev. Lett.* **82**, 4831–4.

[275] Stanislas, M. private communication.

[276] Stromberg, J. L., Mclaughlin, D. K., and Troutt, T. R., 1980, Flow field and acoustic properties of a Mach number 0.9 jet at low Reynolds number, *J. Sound Vib.* **72**, 159–76.

[277] Tafti, D. K. and Vanka, S. P., 1991, A numerical study of the effects of spanwise rotation on turbulent channel flow, *Phys. Fluids A* **3**(4), 642–56.

[278] Tennekes, H. and Lumley, J. L., 1972, *A First Course in Turbulence*, MIT Press.

[279] Thompson, K. W., 1987, Time dependent boundary conditions for hyperbolic systems, *J. Comput. Phys.* **68**, 506–17.

[280] Touil, H., Bertoglio, J. P., and Shao, L., 2002, The decay of turbulence in a bounded domain, *J. Turbulence* **3**, 049.

[281] Tracy, M. B. and Plentovich, E. B., 1997, Cavity unsteady-pressure measurements at subsonic and transonic speed, NASA Technical Paper 3669.

[282] Tseng, Y. and Ferziger, J., 2001, Effects of coastal geometry and the formation of cyclonic/anti-cyclonic eddies on turbulent mixing in upwelling simulation, *J. Turbulence* **2**, 014.

[283] Urban, W. D. and Mungal, M. G., 1998, A PIV study of compressible shear layers, in *9th International Symposium on Applications of Laser Techniques to Fluid Mechanics, Lisbon, Portugal*.

[284] Urbin, G., 1998, Étude numérique par simulation des grandes échelles de la transition à la turbulence dans les jets, PhD thesis, Grenoble.

[285] Urbin, G. and Métais, O., 1997, Large-eddy simulation of three-dimensional spatially-developing round jets, in *Direct and Large-Eddy Simulation II*, J. P. Chollet, L. Kleiser, and P. R. Voke (eds.), Kluwer, pp. 35–46.

[286] Urbin, G., Brun, C., and Métais, O., 1997, Large-eddy simulations of three-dimensional spatially evolving round jets, in *11th Symposium on Turbulent Shear Flows*, Grenoble, September 8–11, pp. 25-23–25.28.

[287] Van Mieghen, J. and Dufour, L., 1948, Thermodynamique de l'atmosphère, *Mem. Inst. R. Met. Belgique*, **30**.

[288] Vincent, A. and Ménéguzzi, M., 1991, The spatial structure and statistical properties of homogeneous turbulence, *J. Fluid Mech.* **225**, 1–20.

[289] Vreman, B., Geurts, B., and Kuerten, H., 1997, Large-eddy simulation of the turbulent mixing layer, *J. Fluid Mech.* **339**, 357–90.

[290] Weiss, J., 1981, The dynamics of enstrophy transfers in two-dimensional hydrodynamics, La Jolla Institute preprint LJI-TN-121ss. See also *Physica D* **48**, 273–94.

[291] Westphal, R. V., Johnston, J. P., and Eaton, J. K., 1984, Experimental study of flow reattachment in a single-sided sudden expansion, Stanford University, MD 41.

[292] Widnall, S. E., Bliss, D. B., and Tsai, C., 1974, The instability of short waves on vortex ring, *J. Fluid Mech.* **66**(1), 35–47.

[293] Yakhot, V. and Orszag, S., 1986, Renormalization group (RNG) methods for turbulence closure, *J. Sci. Comput.* **1**, 3–52.

[294] Yanase, S., Flores, C., Métais, O., and Riley, J., 1993, Rotating free-shear flows. I. Linear stability analysis, *Phys. Fluids* **5**, 2725–37.

[295] Yoshizawa Y., 1986, Statistical theory for compressible turbulent shear flows, with the application to subgrid modeling, *Phys. Fluids* **29**, 2152–64.

[296] Zaman, K. B. M. Q., Reeder, M. F., and Samimy, M., 1994, Control of an axisymmetric jet using vortex generators, *Phys. Fluids* **6**(2), 778–94.

[297] Zang, Y., Street, R. L., and Koseff J. R., 1993, A dynamic mixed subgrid scale model and its application to turbulent recirculating flows, *Phys. Fluids A* **5**(12), 3186–96.